THE STEPS OF CREATION
PART III: EARTH HISTORY

Clare V. Merry

Published by The MerryScience Foundation

Illustrations by Clare Merry

ISBN: 978-1-909204-83-6

The MerryScience Foundation
is an imprint of
Dolman Scott Ltd
www.dolmanscott.com

TABLE OF CONTENTS FOR PART III

Chapter 19 CATASTROPHES STRIKE THE EARTH

Chapter 20 THE ORGANIC ORIGIN OF ROCK STRATA

Chapter 21 THE MECHANICS OF THE EARTH

Chapter 22 GEODYNAMOS

CONCLUSION III

APPENDICES

LIST OF ILLUSTRATIONS FOR PART III

The computer graphics, ink and pencil drawings in this book are by the author. A few have been adapted from other sources where indicated.

Tables have been compiled by the author using data from various sources.

COVER DESIGNS

The pastel drawings by the author for the covers of Parts I, II and III of *The Steps of Creation* represent the early stages of planet Earth when it was inhabited only by microbial life.

The cover of Part I shows the Earth inoculated with chemolithotrophic methanogen Archaea. The Earth is in the darkness of outer space when these microbes start to produce water through biochemistry upon the dry basalt rock surface.

On the cover of Part II Purple bacteria bloom in pools of water on Earth's cratered surface. These anaerobes capture rays from the dim juvenile Sun practicing anoxygenic photosynthesis, and thereby increase the amount of water on Earth. The surface is strewn with the blackened products of anaerobic metabolism and elemental sulphur.

The cover of Part III depicts the dawning of Earth's aerobic atmosphere as the Earth is colonized by cyanobacteria whose photosynthesis releases free oxygen. The shallow seas are lined with bacterial mats and the bare rock surface is colonized – everywhere is green, tinged with the red of rusted iron.

Scenarios of the early stages of planet Earth are given in the following chapters of *The Steps of Creation*:

The stages of planet Earth are described in Chapter 10, section 10, of Part I.

Candidates for life in space are given in Chapter 11, section 7, of Part II.

The stages of proto-planet Earth are described in Chapter 18, section 4, of Part III.

THE THEORY OF NANOCREATION AND ENTROPIC EVOLUTION

Manifesto

The central idea of this theory is that God created at the microscopic, nanoscale level, and then allowed macrostructures to evolve naturally. Entities at the nanoscale level include such things as the hydrogen atom, certain elements, organic molecules and DNA. Macrostructures include such things as galaxies, stars, the Earth with its biofriendly environment and our own bodies. The term 'nanocreation' is adapted from the new science of nanotechnology.

Nanotechnology involves the control of the structure of matter on a scale below 100 nanometres. One nanometre is 10^{-9} metres or a millionth of a millimetre. Nanotechnology is performed using ultraviolet and X-ray optical techniques or an electron beam to position groups of atoms or molecules in an engineered material. This human creative activity may serve as an analogy for the creative activity of God.

Genetic engineers now make transgenic mice by inserting genes from humans into the mouse genome for the purposes of research. This feat of molecular manipulation is performed with pre-existing genes. It is proposed that God can equal human creativity, and better; He could make a new gene that would encode a new protein as a component of life.

Entropic Evolution is inspired from the principle behind the Second Law of Thermodynamics. An increase in entropy is a move away from order towards a system characterized by disorder and inertia. The Laws of Thermodynamics relate to energy, whereas the entropy involved here relates to information; this is known as Shannon Entropy since it was first proposed by Claude Shannon in 1948. The term Entropic Evolution is used to convey the idea that, although evolution gives rise to interesting phenomena, it is in itself a winding-down process.

The Theory of Nanocreation and Entropic Evolution has some hypotheses which seek to define primary causes – the intervention of God in the universe at specific points in time and space, while other hypotheses seek to explain secondary causes – the working of natural evolutionary processes, and their limits. The combination of these primary and secondary causes gives rise to a global scheme. It is unashamedly theistic science.

Nanocreation does not involve a literal reading of the first chapters of Genesis. The theory accords with the essence of the creation account of Genesis, and its underlying truths.
The essence of the Genesis story is that God is the Creator of both matter and life. The days of creation convey the idea that creation occurred as a sequence of acts of creation over time, culminating in the creation of humankind.

The Theory of Nanocreation and Entropic Evolution may be accepted as a scientific theory for the following reasons:

- It is based on data generated by modern, orthodox science and thus shares common ground with other current scientific theories.
- It is a theory composed of hypotheses, not just a literal interpretation of the Bible.
- It has generated predictions that are testable by science, and which lay it open to being proved wrong in part or as a whole; it is a falsifiable theory.

The theory incorporates an explosive beginning to the universe which I have named the Multi-Bang. The Multi-Bang has some important differences with the Big Bang Theory. Energy in the form of motion in created matter gave rise to an expanding universe.

In line with main stream science, the universe is accepted to be around 13.7 thousand million years old. The timescale for the appearance of the first life forms on Earth is counted in thousands of millions of years. It is accepted that radiometric dating places some of the earliest microbial fossils – stromatolites from the Precambrian era at 3.5 thousand million years old.

Evolution is viewed as something positive; its possibility incorporated in the original set-up. Evolution is natural in that it consists of unguided processes happening by chance. There is both cosmic evolution involving the 'heavenly bodies' and biological evolution occurring in the different forms of life.

Charles Darwin presented his Theory of Descent with Modification through Natural Selection in *On the Origin of Species* in 1859. The concept of the modification of pre-existing traits through Natural Selection giving rise to diversity in the organic world is fully incorporated into the Theory of Nanocreation and Entropic Evolution. Darwin mainly establishes the Theory of Common Descent to class level; this is also endorsed by the Theory of Nanocreation and Entropic Evolution. Another meaning of the term Natural Selection expanded by Darwin in the second half of *The Origin of Species* is the step-by-step building-up of new structures and organs representing an increase in complexity, and couched in terms of usefulness. This strand of evolutionary thought is rejected by the Theory of Nanocreation and Entropic Evolution.

The concept of Entropic Evolution in the biological world rests on the well-documented observation that mutation causes dysfunction in the genetic system. Mutation tends to switch genes off such that they are not expressed. However, what is dysfunction in the genome often manifests itself in the organism as modification and variation. Where variation proves useful, it is preserved by Natural Selection. Many examples of evolution, including the classic examples, in fact, involve loss of structures and traits in the organism. Thus, the selection of variations does not build up complexity as Darwin claimed, but it does produce adaptation to the environment and a great diversity of species.

NeoDarwinism emerged in the 20th century by combining Darwinist theory with the laws of inheritance, and the discovery of genes and DNA. A new idea was added – that of the origin of life from non-life by natural processes involving selection. A new philosophy came to dominate science asserting the incompatibility of science and religion.

Mainstream science today has an ethos of rationalism and objectivity. However, many materialistic hypotheses offered within the context of science to answer the big questions are a matter of belief, not of scientific demonstration. They are deemed worthy of science textbooks purely on the basis that they do not involve any interaction between a Creator and creation; it is thought that their materialism makes them 'scientific'. Science today follows the Philosophy of Naturalism – a philosophy unknown to the founders of modern science. Naturalism accepts only natural explanation for the working *and* origin of all phenomena, and excludes God from its definition of reality. This *a priori* stance does not make the many scientific theories necessarily objective, nor even rational.

The Theory of Nanocreation and Entropic Evolution is a rejection of purely materialistic science and the Philosophy of Naturalism. It is science which does not exclude God. To the debate that rages between Creationists and Evolutionists, the Theory of Nanocreation and Entropic Evolution offers a middle way: creation in the context of an evolving world.

Introduction

Parts I and II of *The Steps of Creation* are about the Theory of Nanocreation and Entropic Evolution. Part III contains hypotheses relating to natural processes of wide application and especially Earth history.

In *The Steps of Creation* I will present the current theories and hypotheses of science – that is, the interpretation of the facts which are given in science textbooks, popular science programmes and encyclopedias. I will present my alternative theory and hypotheses as a different interpretation of the same facts. I will also discuss some ideas, notions and views that are either religious or philosophical, and which relate to science and the natural world.

The Theory of Nanocreation and Entropic Evolution is composed of nine hypotheses; six hypotheses are presented in Part I and three in Part II. There are three additional hypotheses in Part III which are not part of the Theory. *The Steps of Creation* is presented in three volumes corresponding to the three parts: Part I Life; Part II Cosmos; and Part III Earth.

The hypotheses in Part I relate to Nanocreation and Entropic Evolution. These hypotheses concern the DNA code and genetic system, the classification of unicellular life, plants and animals, the origin of life, the formation of ecosystems, and the origin of water and the conditions for life.

The hypotheses in Part II relate to Nanocreation and cosmic evolution. The hypotheses of these chapters relate to the origin of the elements and extraterrestrial life, the formation of stars, subatomic particles, the explosive beginning to the universe, and the formation of galaxies.

The hypotheses in Part III concern Earth history. The hypotheses of this last part propose mechanisms that are natural processes bringing about the formation of the solar system and Earth as a planet, and moulding the features on the surface of the Earth. In Part III Earth history is presented in a way it has never been presented before.

Some readers may accept the whole Theory of Nanocreation and Entropic Evolution, while others may accept some hypotheses and not others. It is my view that we should be free to choose what we believe and free to question so-called 'proven facts'. Science has never advanced without questioning. The Theory is offered in the spirit of debate.

Part I of *The Steps of Creation* covers the following subjects:

- Genetics (also called molecular biology)
- Biology and some human biology
- The classification of organisms (also known as taxonomy or systematics)
- Palaeontology
- Origin of life theories
- Biochemistry
- Microbiology

Part II of *The Steps of Creation* covers the following subjects:

- Astronomy of stars, interstellar clouds and planets
- Astrophysics (the physical and chemical constitution of stars)
- Chemistry
- Physics
- Particle physics
- Cosmology (the origin and development of the universe)

Part III of *The Steps of Creation* covers the following subjects:

- Astronomy (of the solar system)
- Geology
- Earth sciences

A basic understanding of each of these subjects will be given in terms of facts and current theories. At the same time, the hypotheses I present belonging to the Theory of Nanocreation and Entropic Evolution, will show how the facts can be understood in new ways and in relation to the 'Steps of Creation'.

This book will not include details concerning the scientific methods used in investigation or the way in which facts are derived from raw data. There will be no explanations of the workings of scientific instruments. There are many good introductions to science which explain the methods used very well. At most, I will mention the name of the method used in the relevant scientific investigation, only in passing. The reader can use this to look up the methods in an encyclopedia if they wish.

The aim of this book is to get an overview of the facts generally accepted as such by science, and with knowledge of these facts seek further understanding.

One could engage in a long discussion on the theology and philosophy arising from this book, but I do not think it would be profitable. I do not wish to wander too far from the physical details of the natural world.

The English spelling used in this book is from *English Dictionary* Geddes & Grosset (1999) New Lanark, Scotland. The choice of 'z' rather than 's' accords with older English spelling. The use of 'z' is more acceptable to the Americans, and I prefer it.

I will use 'thousands of millions', rather than 'billions'. This is because many people do not know what a billion is. In Britain a billion has been, until recently, a million million, while in the USA it is only a thousand million. The words 'thousand million' make it clear what we are talking about and avoid confusions with zeros.

This book will not be over-burdened with mathematical equations. As scientific words are introduced, they will be explained in the text. Scientific jargon is avoided where possible.

In *The Steps of Creation*, hypotheses, theories and models in science will be given capital letters. My own theory and the hypotheses of which it is composed will also be given capital letters. Likewise, philosophies and some concepts will be capitalized. This is because this book is about ideas and concepts in science. Ideas, notions, concepts and ways of viewing things are to be distinguished from known, proved and agreed-upon facts. A fact will be taken as a fact when there is no longer any dispute as to its existence or veracity.

Normally, Natural Selection is not given capital letters in science books, and although a concept, it is treated as a fact. I have used capitals for Natural Selection to emphasize that it is a concept with various possible meanings. The various possible meanings of Natural Selection are central to the discussion contained in *The Steps of Creation*.

If I have capitalized Natural Selection, should I have capitalized evolution? Is evolution fact or theory?

The word evolution has a much wider application than the term Natural Selection. Sometimes we are all agreed that something has evolved, and so it presents itself as a fact. In other instances, the term evolution competes with other explanations whether religious or philosophical. In this case, evolution presents itself as a belief or theory used to explain nature, and it would be clearer to write it as 'Evolution'. I finally decided to keep evolution without capitals due to the ambiguities involved. I, likewise, decided to leave creation without capitals, although all sorts of discussions about what we mean by 'creation' could be had (I make an exception to this rule in Chapter 5 because this chapter is a discussion of ideas).

In our society evolution is part of our world-view and dominant philosophy. The word 'evolution' or 'evolved' is constantly slipped into common speech, often where it does not belong. A phrase often used is "the evolution of life on Earth", when what is meant is "the existence of life on Earth" since we do not finally know where life came from.

Stars evolve, life evolves and humans have evolved. This may be true, but it still does not explain the origin of these things.

The assumption is that things, in fact, everything around us is the way it is because it evolved that way. It is the unexamined assumption that puts an end to further questions. This is the world in which we live and which we must examine – if we can just stand back from it a bit.

The Steps of Creation is a book about agreed facts, and debatable theories and concepts. It offers new scientific interpretations of the facts. It presents a reality behind which is God – Creator, Sustainer and Father of all.

Natural Theology, traditionally seeks to prove the existence of God through the observation of nature. This has been done through expounding the design found in nature, and the sense of balance and harmony in the natural order. I am not trying to prove that God exists. I am sharing my views on creation with fellow believers who believe in God for their own personal reasons, and because of their own experiences of spiritual things, and with anyone else who cares to listen.

God is Being, and so requires no justification for His existence other than Himself. Science cannot ultimately prove or disprove the existence of God, since God is beyond the physical realm. However, because God is a Person, He can prove His own existence, as and when, He chooses.

Thus, the existence of God is taken as given as regards the Theory of Nanocreation and Entropic Evolution.
I seek an understanding of nature as a search for truth and a valid enterprise in its own right, not to prove to anyone that God exists. I believe that God does not wish us to prove His existence with science, because if He had wanted us to do so, He would have made it easier.

My objective is to explain nature, but my scheme only makes sense if there is a Creator God. To Him be the glory.

CHAPTER 17

THE SOLAR SYSTEM AND PLANET CAPTURE

Introduction

This chapter focuses on the solar system and especially the motion of planets. I contest the Nebular Theory which is the current explanation given for the formation of the solar system. By doing this I break with a two hundred year tradition in science.

The Nebular Theory describes the solar system as a self-contained system. I present the solar system as being formed of diverse bodies brought together by the gravitational attraction of the Sun. The principle of capture that I will present is a universal principle. It applies to all the heavenly bodies in the universe, whether galaxies, stars, stellar corpses, brown dwarfs or planets. It also applies to asteroids, meteorites and comets.

1. The Nebular Theory

The Nebular Theory is a model of how the solar system might have formed. The theory goes right back to the Marquis de Laplace, a French astronomer of the 18th century. In 1796 Laplace proposed that the solar system originated as a rotating disk of gas and dust that he called the solar nebula. As the nebula cooled and contracted, the planets condensed out of rings of gas with the Sun forming in the middle.

The model reflected the general regularities observed in the solar system; that the planets orbit the Sun in the same direction (anticlockwise when viewed from the North Pole), they orbit generally in the same plane, they generally rotate in the same direction and are regularly spaced. The theory has survived to this day. However, observation has shown increasingly that the solar system has many curious irregularities in orbits and rotations that contradict the model.

Planetesimal Hypothesis

Simulations have shown that planets could not simply condense out of the original nebula as the classic theory required. The Nebular Theory was therefore modified with the Planetesimal Hypothesis. The Planetesimal Hypothesis proposes that planets are built up by accreting material piece by piece through collisions. The hypothesis still starts with dust in the original disk. The dust is thought to have formed grains that then clumped into metre-sized boulders. The boulders coalesced into larger masses and these larger masses collided to form planets. These larger masses are the planetesimals. Simulation shows that planets can only form from the coalescing of these large object planetesimals, not from fine dust. The planetesimals grew into the terrestrial planets and the cores of the giant planets.

The giant gaseous planets must, according to the theory, have acquired their hydrogen and helium from the original solar nebula gas early on before it disappeared, and so the planet formation process must have occurred early on (Taylor 1998, page 59). Although Taylor also points out that Jupiter does not have the composition of the Sun, as one would expect if it was simply a piece of the original nebula.

Thus, it is hypothesized that micron-sized dust particles aggregated into planetesimals of one metre to ten kilometres in diameter (Shaw 2006, page 159). But no adequate explanation has been given to show why small grains would clump together to form objects instead of simply dispersing. Small-sized particles have tiny spheres of influence that are not sufficient to attract and hold other matter by gravitation. The planetesimals are then supposed to undergo further aggregation by collision. However, when objects collide in space, they tend to fragment rather than merge together. I see problems with the initiation of planet formation, although I agree that objects above a certain threshold with a bigger gravitational sphere of influence could attract matter and coalesce into still larger objects, eventually forming planets by accretion.

The solar nebula disk and angular momentum

The solar system as it is now is quite thin. If it were scaled to the size of a small pizza, it would be one centimetre thick, with the orbits of the planets contained within this thickness. The solar nebula is supposed to have once resembled a large rotating pizza base in space made from gases. A disk of gas does not exist now. According to the theory most of the nebula finished up in the Sun.

The Sun contains 99.9 % of the mass of the solar system. There is no evidence now that the Sun's material came from a disk-shaped solar nebula. A major problem with the Nebular Theory is the observed distribution of angular momentum in the solar system. The planets are orbiting the Sun much faster than the Sun itself is rotating. If the planets and the central Sun had indeed formed from a rotating cloud, one would expect that as the cloud contracted and in accord with the conservation of angular momentum, that the central object would be spinning faster and faster. In fact, although the Sun holds over 99 % of the mass of the entire system, it has less than 1 % of the angular momentum. If the Sun had 99 % of the angular momentum, it would spin once every few hours, instead of once every 25 days.

Various ad hoc hypotheses have been added to try to resolve the angular momentum problem. It has been proposed that the protosun was originally spinning much faster, with the dust cloud revolving around the Sun going correspondingly more slowly; by some mechanism spin was transferred from the Sun to the dust cloud, thus slowing the former and speeding up the latter. However, using get-out clauses to get around the laws of physics is not proving very satisfactory.

2. Planet Capture Hypothesis

I propose that the solar system is not a self-contained system as the Nebular Theory proposes. I propose that the solar system of planets orbiting the Sun was formed principally by the fortuitous capture by the Sun of planets that were travelling through space along with various other objects.

In 1994 I had been studying what is known about the solar system as an off-shoot in my search to understand the origin of life on Earth. I concluded that the planets do not appear to have the same age or origin.

Later that year I returned to reading up about the extinctions of life on Earth. I came across the hypothesis proposed by Luis Alvarez in the early 1980s which attributes the extinction of the dinosaurs to an asteroid hit to the Earth. This set me thinking; why would not all five major extinctions on Earth be attributed to events in the solar system? So I formulated the idea that planets younger than the Earth might be captured into the solar system by the Sun and upon arrival disrupt the Earth, causing extinctions (I have since modified my original idea, but kept the same line of thought).

I knew about supernovas and the power of explosions to set objects in motion –thus in my mind I could see all sorts of explosions occurring in the galaxy and all sorts of objects flying through space waiting to be captured. The number of extinctions on Earth and the number of possibly younger planets seemed to tally. Thus, I could hardly contain my excitement, although I have contained it for 15 years! (I am going to return to mass extinctions on Earth in Chapter 19).

The **Planet Capture Hypothesis** can be formulated thus:

The solar system was formed by our star, the Sun capturing travelling proto-planets through gravitational attraction. Planets originated as molten iron planetary cores forged within red giant stars and shot out by nova or supernova explosions. These molten iron cores accreted silicate mantles of cold interstellar dust while travelling through space. Thus, the planets of the solar system are of different ages and origins.

The Sun, like other stars, would have formed in a molecular cloud within an eddy formed of molecular hydrogen gas. The formation of stars was discussed in Chapter 13 of Part II. The angular momentum of the Sun would have come from the rotation of the molecular cloud. The axis of rotation causes rotating systems to assume a disk shape; this explains the plane of the solar system.

Planetary cores and proto-planets
Dying stars would have drawn in iron, silicates and carbon molecules from surrounding dust clouds in their red giant phase. The silicates would melt into lumps of rock, while the iron became smelted into globules of molten iron within the star. At a certain point in time, explosion would be triggered by instability possibly produced by the transformation of some elements into radioactive elements inside the star or by water ice drawn into the star changing suddenly into steam.

The globules of molten iron inside the star would be shot out by the nova or supernova explosion along with debris of various types and sizes.

The molten iron globules would, in effect, be hot denuded planetary cores travelling through the galaxy each along its own path. The silicate debris would form meteorites also travelling outwards and possibly fragmenting along the way.

Over time high numbers of nova and supernova explosions of red giant stars would have taken place launching planetary cores and all grades of debris into space. The planetary cores would travel in all directions, criss-crossing the galaxy. This larger graded matter cannot be detected by telescopes because it does not emit light nor interact with light.

The molten iron planetary cores, as they move through interstellar dust clouds and debris clouds left by red giant novas, would accrete rocky material. The rocky material would form a mantle around the molten core, turning the planetary core into a proto-planet (this will be discussed further in Chapter 18, section 2).

If the travelling proto-planet came into the vicinity of a star, it may be captured by gravitational attraction. In certain conditions the proto-planet would take up an orbit around the star, accreting matter in the vicinity of the star and becoming a planet.

Our solar system would have been formed by the capture of planets over time. These planets would have diverse origins reflected today in their diverse compositions and different speeds of rotation.

The momentum of the planets would come from the nova or supernova explosion which originally set them in motion. Planets would travel in straight lines, but they could become captured in the curved geometry of space around large bodies such as stars. The orbits of the planets became ordered by various forces, to be examined shortly. The rotation of planets would also have been set off by the original stellar explosion. Thus, the angular momentum of the solar system would come from explosive events outside the solar system, not from within the solar system by contraction of a disk-shaped cloud.

The evidence in support of a supernova origin to planetary cores (and meteorite bodies) is firstly that nickel-56, cobalt-56 and iron-56 have been associated with supernovae. Secondly, planetary cores are thought to be formed of iron and nickel. Thirdly, alloys of iron and nickel occur in nickel-rich meteorites, often with impurities of cobalt and carbon. The world's nickel resources come from meteorite impact sites such as the Sudbury Basin in Canada. Therefore, there is a connection between supernova explosions, metal-rich bodies travelling through space and planetary cores.

Another line of evidence supporting a supernova origin of the Earth is the fact that the Earth's rotation is slowing down. Six hundred million years ago there were over 420 days in a year with each day lasting 21 hours. The Earth's rotation is slowing down by about 1.4 milliseconds a century (Ince 2007, page 40). If the Earth had been set spinning as a planetary core launched by an exploding star, the gradual slow-down in rotation is what one would expect.

The Jovian Planets
We established in Chapter 13 that the giant gaseous planets show the features of brown dwarf stars. They are smaller and cooler than other known brown dwarfs probably because they are older and so have cooled down and shrunk in size. This means they can be seen as small failed stars. The Jovian planets (their other name) are orbited by terrestrial satellites and rings of dust, ice and rubble. They resemble tiny solar systems. One possibility is that the Jovian planet with its satellites and rings was captured by the Sun as a whole unit. The other possibility is that small incoming terrestrial bodies as they passed through the outer solar system towards the Sun strayed too near to a gaseous giant and were caught into orbit around it. Both these possibilities may be true since they are not mutually exclusive; planets may be captured into the solar system with orbiting satellites or moons, and then in their turn capture more incoming bodies.

Therefore, I propose that the Sun and planets do not have a common origin. There was never any solar nebula disk of gas. Planets do not simply condense out of gas or form by the aggregation of dust particles into planetesimals with no starting point. The elements of which the planets are composed do not simply form or amass themselves in certain parts of the solar system according to distance from the Sun.

14

The hypothesis I am proposing departs radically from the currently held Nebular Theory.

I have made a connection between nova and supernova explosions, objects travelling randomly through space, gravitational capture by bodies of high mass such as stars, the fortuitous formation of the solar system, and extinctions on Earth.

Since then I have become aware that the role of chance events in the formation of the planets, the moon and satellites (satellite is another name for moon) has become very popular. Some of the new ideas give backing to my own ideas. None has gone so far as throwing out the Nebular Theory as I have. The popular hypotheses today keep on adding new modifications to the accepted theory.

Hypotheses emphasizing chance events

If there once existed a solar nebula that came to form the Sun and planets, one would expect that the planets would either have the same composition or exhibit a regular systematic change in composition with distance from the Sun. The problem is that the planets are all different. It is true that there are terrestrial planets in the inner solar system, giant gaseous planets further out and large icy planets beyond them. However, Pluto is a small terrestrial but icy planet on the outer boundary of the solar system (although one solution to this problem has been to reclassify Pluto as not being a planet).

One would also expect the planets to have vertical axes and rotate at the same speed or a speed dependent upon distance from the Sun "in some tidy mathematical sequence" as Taylor (1998, page 169) remarks. However, all the planets are tilted on their axes, and all rotate at different speeds. Venus even rotates backwards and Uranus lies on its side.

Stuart Ross Taylor in *Destiny or Chance: our Solar System and its Place in the Cosmos* (1998) proposes that chance events have shaped the course of the history of our solar system. These chance events involve dramatic collisions of planets with massive bodies and some capturing of satellites by the giant planets.

According to Taylor the untidy situation of the planets has been caused by massive collisions which occurred in the final stages of

15

planet formation causing the planets to become tilted and start spinning at different rates. A giant impact with the Earth splashed out the Moon and caused the Earth to start rotating. Another impact stripped Mercury of its rocky mantle. Collisions caused disks to form around the giant planets from which satellites formed. Uranus was knocked onto its side by an impactor the size of the Earth. However, the slight tilt of Jupiter and the 30° tilt of Saturn he explains by warps in the gaseous solar nebula (Taylor 1998, page 171).

Taylor proposes that the impactors were large bodies that existed in the solar system, but which have since vanished. The evidence for this, he claims, is the observation that the surfaces of planets and satellites are saturated with craters. It is proposed that there was a period of intense bombardment in the solar system which mainly ended by 3850 million years ago.

The regular satellites around the giant planets are described as having formed like miniature solar systems from disks around these planets. But for the irregular satellites Taylor offers a different explanation; they are captured bodies that were bits and pieces that were not swept up by planets as they grew. An example of an irregular satellite is Triton which is orbiting Neptune in a backwards direction with a high inclination to the plane. Taylor likens the arrival of Triton to a "bull entering a china shop" and destroying by collisions any original satellite system around Neptune (Taylor 1998, page 87).

Thus, Taylor offers a science of chance events in which each specific detail is attributed to a random cause. In this scenario many random events occurred during the history of the solar system resulting in planets and satellites that are all different. I welcome the emphasis on the observation that the planets are all different. However, the hypothesis that I present here attempts an understanding of the interaction between gravity and motion in planetary orbits, and speed of rotation about the axis that will reveal regularities in the universe. If regularities can be established, predictions about planets should be possible.

Taylor asserts that the problem satellites –those that orbit the gaseous giants in retrograde or highly inclined orbits –have been captured by these planets. He asserts that previous to capture these satellites consisted of bodies moving aimlessly through the solar system having been formed from the nebula, but not having found an orbit. The planets and their satellites that have prograde orbits would have been formed from the nebula according to the orthodox Nebular Theory. Planets with strange tilts such as Uranus lying on its side, he claims would have been knocked into this position by a massive collision. Thus, Taylor adopts a different hypothesis for each case. Just as new things can be discovered about planets, additional hypotheses can be invented to cover the new cases.

I came across Taylor's ideas 15 years after formulating by own ideas about the origin of the solar system. I find it encouraging that someone else is exploring along the same track as I. However, what I propose is a single theory that covers all cases. Totally clearing away the clouds of the Nebular Theory, my theory is solely based on capture.

Roaming proto-planets

To recap, the vaporization of water and the accumulation of radioactive elements in a red giant star eventually destabilize the star and it explodes like an atomic bomb. Material of all sizes is shot out by a nova or supernova. Molten iron lumps are shot out like bullets. These lumps assume a spherical shape because they are molten and rotating, and they remain as one lump due to the cohesive properties of iron. Silicate dust has been melted into rocky meteorites of all different sizes that are also shot out of the red giant. Rock and rock-iron conglomerates have a tendency to break up into smaller pieces as they travel. This limits the size of meteorites. The supernova also gives rise to a cloud of gases, radioactive elements and dust debris.

Clouds of gas and fine dust will only expand away from the explosion for a certain distance, and then disperse. Objects of higher mass, once they have been given impulse from the force of explosion will start to travel outwards from the centre of explosion through space. Just as a rocket travelling in the vacuum of space and free from strong gravitational forces only needs a small burst of propulsion to keep it moving almost indefinitely, a planetary core will also continue its journey through the galaxy and maybe beyond almost indefinitely.

Each proto-planet continues to follow its own path until it is captured by a star and becomes part of a planetary system. It is possible that the solar system could contain planets originating in another galaxy if one galaxy has passed through another galaxy.

3. The orbits of planets

There are nine planets if Pluto is included as a planet. Today Pluto has been demoted from planetary status since it makes the solar system look messy (see footnote[1]), but for the Planet Capture Hypothesis Pluto's status is irrelevant since any body would be captured by the same forces. Mercury, Venus, Earth, Mars and Pluto are small terrestrial planets, while Jupiter, Saturn, Uranus and Neptune are giant gaseous planets. Many of the satellites orbiting the gas giants resemble the smaller sized terrestrial planets.

Before proceeding any further I am going to define the units of measurement used in descriptions of the solar system. The astronomical unit or AU is the distance between the centre of the Sun and the Earth. It is about 150 million kilometres or 93 million miles. A light-year is about 63 000 AU. The nearest star, Proxima Centauri, is 4.3 light-years away, or 270 000 AU.

Planetary motion and gravitational attraction

Galileo Galilei (1564-1642), Italian astronomer proposed the principle of inertia in 1640: if an object is left alone, and no forces are applied to it, then its state of motion is unchanged. If it is moving, it continues to move at the same speed and in the same direction. If it is standing still it will continue to do so.

Sir Isaac Newton (1642-1727) English physicist and mathematician published *Philosophiae Naturalis Principia Mathematica* in 1687 in which he formulated three Laws of Motion which became the basis to modern physics (see footnote[2]).

---

[1] In 2006 the International Astronomical Union demoted Pluto from the status of planet to dwarf planet, along with Charon (Pluto's satellite), Eris (once called UB313 and marginally bigger than Pluto), and Ceres (the biggest asteroid in the asteroid belt).

[2] Isaac Newton's three laws of motion are as follows:
1. A body remains in its state of rest or motion unless it is compelled to change that state by a force impressed upon it.
2. The change of motion which is the change in velocity times the mass of the body, is proportional to the force impressed upon it.
3. To every action there is an equal and opposite reaction.

Newton's First Law of Motion is based on the principle of inertia. It states, 'Bodies move in a straight line with a uniform speed or remain stationary unless a force acts to change their speed or direction.' The conservation of linear momentum is defined as the product of the mass of the body and the velocity. It describes the difficulty in slowing down a moving body; a heavy body moving quickly has high momentum and is difficult to deflect or stop. It is also difficult to get a heavy body moving if it is stationary.

In fact, there is no such thing as perpetual motion since friction causes a force which decelerates a moving body. Newton's First Law can only be seen as a special case where there is no friction. The nearest situation to this is that of bodies moving through space where the density of atoms is very low.

In the context of the Planet Capture Hypothesis, a planetary core will travel further than fine dust particles due to the greater mass of the planetary core allowing the conservation of linear momentum. The planetary core would continue to travel through galactic space on the basic principle of inertia almost unimpeded by friction since atoms are so sparse in interstellar space. The planetary core will conserve the momentum of its trajectory and the angular momentum of rotation on its axis.

Newton hit upon the idea of the force of gravity when struck upon the head by the proverbial apple, although skeptics hotly dispute the existence of this apple. Whatever the true story, Newton introduced the Law of Universal Gravitation in the *Principia* and this allowed him to calculate the motions of planets in the solar system. Gravity is an attractive force between massive bodies which does not require bodily contact and which acts at a distance. The gravitational force between bodies depends on the masses of the bodies and is measured from the centre of the body. Newton deduced that the gravitational force between bodies diminishes as the inverse square of the distance between the bodies. Thus, if the distance between the bodies is doubled, the force on them is reduced to one-quarter.

The effect of gravity means that a body has a sphere of influence depending on its mass in which it can hold other objects. A body with a large mass has a large sphere of influence. The sphere of influence of the Sun holds the planets in orbit. The gravitational force being inversely proportional to the square of the distance between two bodies means that it drops off rapidly with distance.

The escape speed is the minimum speed a body needs to attain to be free of the gravitational bonds of a larger body and break out of its sphere of influence. The escape speed is higher as the mass increases or the radius decreases of the body trying to escape. The escape speed is directly proportional to the square root of the mass and inversely proportional to the square root of the radius (Zeilik 2002, page 86). To escape from Earth's gravity a rocket needs to reach an escape velocity of over 11 km/second (about 25 000 miles per hour).

Robert Hooke (1635-1703), English physicist and contemporary of Isaac Newton with whom he had many disputes, described planetary motion as the continuous diversion of a rectilinear motion by a central attraction. On this point Newton was in agreement with Hooke since in Newtonian physics planetary motion continues due to inertia while the orbit around the Sun is maintained by the force of gravity.

Newton was a Unitarian – the Unitarians are an off-shoot from Reformed Christianity. Unitarians reject belief in the Trinity. Newton believed that the planets were set in motion and placed in their orbits around the Sun by God. The force of gravity was the mechanism which kept the clockwork of the solar system running. Newton's calculations showed that periodic adjustments were necessary to maintain the planets in their orbits. He believed that God made these periodic adjustments like a clockmaker winding up a clock such that the solar system could continue to exist for an infinite amount of time.

With the Planet Capture Hypothesis I am proposing that the motion of planets came from the nova and supernova explosions which gave birth to them, while gravity has drawn them into the solar system. The solar system will not have an eternal duration. I am not proposing that God created the solar system directly, although I do believe that God created matter with properties that would define gravity.

The encyclopedia entry for gravitation states that the "trajectories of bodies in the solar system are determined by the laws of gravity" and that "Until Newton's findings, it was not realized that both the movement of celestial bodies and the free fall of objects on Earth are determined by the same force." (*Encyclopedia Britannica 2011*

Standard Edition: Gravitation). It appears to be generally believed today that the force of gravity accounts for everything or if some angular momentum needs to be added in, it is assumed that the original solar nebula disk had angular momentum and the difficulties with this as noted in the first section are not dwelled upon. Newton and his contemporaries did not view gravity in this all-encompassing manner.

Gravitational theory was further developed by Albert Einstein (1879-1955) in the 20th century with his Theory of General Relativity. The major contribution of Einstein's theory is in its radical conceptual departure from classical theory. In the modern field theory of General Relativity the acceleration due to gravity is a purely geometric consequence of the properties of spacetime in the vicinity of attracting masses. Gravity is the curvature of spacetime in the presence of mass.

The Planet Capture Hypothesis accords with an Einsteinian understanding of gravity in that a planetary core moving through space travels in what appears to be a straight line. If it approaches a star whose mass distorts the geometry of spacetime, the planetary core continues its trajectory caught within the curvature of spacetime. The extreme curvature of space may capture the proto-planet into orbit around the star. Thus, the orbit of the planet is a continuation of it travelling in what appeared to be a straight line. The orbit of a planet becomes conformed to the solar system norm by other forces which produce regularities.

Centripetal and centrifugal forces
Have you ever watched the water from your shower or when emptying the bath and wondered why the water always goes down the plug hole in an anticlockwise direction in the northern hemisphere, and a clockwise direction in the southern hemisphere? Even when you swoosh it in the opposite direction, the water still reverts to the direction it wants to go in. The reason for the motion of the water is that the water follows the direction of rotation of the Earth.

Just as water can approach a plug hole from any direction and assume the motion of the Earth, a proto-planet approaching the solar system will be drawn in with conformity to the direction of motion of the system. A centrifugal or centripetal force is motion about an axis. The reason why the planets all orbit the Sun in the same direction is that they were drawn into the system in a common direction according to a centrifugal or centripetal force.

Plane of the ecliptic

The planets orbit the Sun close to one plane, the plane of the ecliptic. The plane of the ecliptic is defined by the orbit of the Earth around the Sun. The solar system is disk-shaped – like a pizza base in space. The inclination of a planet is the angle that the orbit of the planet makes to the plane of the ecliptic. The tilt or obliquity of a planet is the angle of the rotation axis of the planet relative to the plane of the ecliptic.

According to the Nebular Theory, the solar nebula turned into a disk due to centrifugal forces and the planets formed within this disk and hence all lie close to one plane. According to the Planet Capture Hypothesis, proto-planets approach the solar system from different directions but they are drawn into the plane of the solar system. After a certain number of orbits of the Sun, they conform ever closer to the plane. The solar system has a plane because it is rotating on an axis. The rotation of the Sun would have originated from the rotation of the molecular cloud which gave birth to it. An eddy giving rise to a protostar would be subject to centripetal forces – a force drawing matter in. In the Planet Capture scenario matter continues to be drawn into the Sun. In standard astronomy the solar system is subject to a centrifugal force throwing matter out, although the measurements of angular momentum of the planets do not support the model.

The plane of the ecliptic is defined by the Earth's orbit around the Sun only for historic and Earth-centric reasons. It would make more sense to define the plane of the ecliptic by the orbit of the largest body in the solar system and the oldest body. A large body will be stable and an old body will have had longer to conform to the plane set by the original conditions. The plane of the ecliptic should be set by Jupiter's orbit around the Sun. Jupiter has 318 times the mass of the Earth. The inclination of the orbit of Jupiter to the plane of the ecliptic is 1.3 degrees. Taking Jupiter as the standard would change the inclination and tilt of other planets to some extent.

After a great number of revolutions around the Sun, each planet will align itself on the plane of the solar system disk. However, more recently arrived planets may still be in the process of aligning themselves. There are three planets that show considerable inclination to the plane of the ecliptic: Venus has an inclination of 3.4 degrees, Mercury has an inclination of over 7 degrees, and Pluto has an inclination of 17.2 degrees. This would indicate that Venus, Mercury and Pluto showing the least conformity to the plane, are the most recent arrivals in the solar system.

Of the other planets, Mars has an inclination of 1.85 degrees to the plane of the ecliptic. Neptune's orbit is inclined at 1.8 degrees, Uranus' orbit is inclined at 0.8 degrees and Saturn's orbit is inclined at 2.5 degrees.

Elliptical and circular orbits

An elliptical orbit is measured by its eccentricity. A perfect circle has zero eccentricity. An extremely elongated elliptical orbit has an eccentricity of 0.999. The planets trace elliptical orbits around the Sun, but most of them are almost circular with eccentricities of less than 0.1. Pluto and Mercury are the exceptions with orbital eccentricities of 0.25 and 0.21 respectively.

In his elaboration of the laws of motion and gravitation, Newton described a thought experiment involving a cannonball fired from a high mountain and aimed parallel to the ground. A cannonball fired with little powder falls to the ground. If more powder is added, the ball travels further along the Earth's curve before falling to the ground. This is one type of elliptical orbit. If the cannon is fired with a large enough charge, the ball goes completely around the Earth in a circular orbit and returns to the cannon. In this example the inertial motion of the cannonball exactly offsets the falling caused by gravity. A specific speed is required to produce a circular orbit. If an even larger charge is placed in the cannon, the greater starting speed allows the cannonball to travel around the Earth in an elliptical orbit. If the starting speed is increased, the elliptical orbit becomes more eccentric. If the speed of the cannonball is fast enough, it will reach escape speed. At escape speed the ball leaves the gravitational grip of the Earth –it travels out and never returns. (This thought experiment is described in Zeilik 2002, pages 84-85).

The principles described by Newton are used when artificial satellites are placed in space. To achieve a nearly circular orbit for a satellite around the Earth, it must be launched in a rocket with several 'burns' at just the right time, in the right direction with the right thrust.

Applying Newton's laws of motion and gravitation to the Planet Capture Hypothesis it can be seen that the orbit of a planet, whether circular or elliptical depends on the velocity of the proto-planet upon capture; the greater the velocity, the more elliptical the orbit. When the orbits of planets and Newton's laws are described in astronomy textbooks, the fact that planets have elliptical orbits is ascribed to Newton's first definition of an elliptical orbit whereby a cannonball falls towards the Earth. In the same way, a planet is thought to fall towards the Sun. In the Planet Capture Hypothesis the planets have elliptical orbits because they resemble the second case of an elliptical orbit in which the cannonball has enough speed to have an elliptical orbit around the central object, but not enough to reach escape speed.

Therefore, the initial conditions influence the form of the orbit. The velocity of the incoming body will determine the type of orbit that it takes up. The planets in our solar system have elliptical orbits due to 'excess velocity' and the greater the velocity, the more eccentric the orbit.

As has already been mentioned, bodies are drawn in by the gravitational attraction of the Sun, however, if they come near to a planet, especially a giant planet, they may be captured by the gravitational sphere of influence of that planet. There are many bodies orbiting the giant gaseous planets, and some bodies orbiting the small terrestrial planets. The same principles apply to all of these cases.

The forms of orbit of planets produced by incoming bodies of varying velocities are as follows:

- A low speed incoming body will result in a planet which spirals into the Sun or a satellite which spirals into a giant planet and is lost.

- A moderate speed proto-planet may be captured into an almost circular orbit.

- A high speed proto-planet will be captured into an elliptical orbit; the faster the initial speed, the more eccentric the orbit.

- A very high speed body will have enough speed to equal the escape speed, and so it will enter the solar system and exit again without being captured.

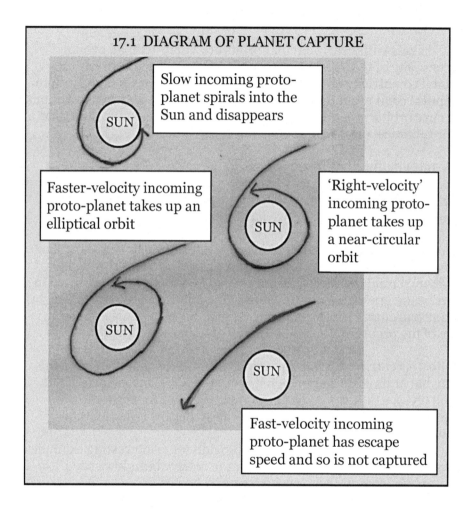

17.1 DIAGRAM OF PLANET CAPTURE

Slow incoming proto-planet spirals into the Sun and disappears

Faster-velocity incoming proto-planet takes up an elliptical orbit

'Right-velocity' incoming proto-planet takes up a near-circular orbit

Fast-velocity incoming proto-planet has escape speed and so is not captured

Slower moving bodies may be captured with more ease by a giant planet and thus remain orbiting the giant planet in the outer solar system. The planets of the inner solar system must have enough momentum to resist the pull of gravity of the Sun and remain in orbit or they would be drawn in and disappear.

There may be planets which, lacking in momentum spiralled into the Sun and have been absorbed leaving no recognizable trace of their passage. The gaseous giants may also have gobbled up some satellites. Also any proto-planet visitor travelling at escape speed will have disappeared out of the solar system leaving no recognizable trace.

The planets with nearly circular orbits are Venus with an orbital eccentricity of 0.007 and Neptune with 0.009. The Earth has an orbital eccentricity of only 0.017, while the Moon has a slightly more elliptical orbit around the Earth of 0.055. Jupiter, Saturn and Uranus all have orbital eccentricities that can be rounded to 0.05. Mars has a more elliptical orbit with 0.093.

The most elliptical orbits of planets in the solar system are those of Mercury with an eccentricity of 0.206 and Pluto with an eccentricity of 0.246. These two planets also have the most inclined orbits to the plane of the solar system as we have seen.

The highly eccentric orbit of Mercury is interesting in that the major axis of the orbit shifts around the Sun. The orbit turns through 360° in about 3 million years. The advance of Mercury's orbit is caused in part by the gravitational attraction of the other planets, but mostly by the strong curvature of spacetime close to the Sun. This is taken as one of the proofs of Einstein's Theory of General Relativity.

Pluto's eccentric orbit has an average distance of 39.44 AU from the Sun, but at its closest approach it is at 28.7 AU. This means that its orbit crosses the orbit of Neptune which is at 30 AU from the Sun. Pluto and Neptune do not collide because they orbit in different planes.
The satellites of the gaseous giants provide some interesting examples of orbits. There are examples of what may have been low speed captured bodies and high speed captured bodies.

The largest satellite of Neptune, Triton has retrograde rotation and a retrograde orbit; it is orbiting backwards around the planet in a highly inclined orbit. The orbit is nearly circular, but Triton is spiralling in towards Neptune, and will eventually collide with it. These facts indicate the recent capture of a low-velocity body. The other major satellite of Neptune, Nereid has the most elliptical orbit of all satellites in the solar system. This indicates the capture of a high-velocity body.

If some satellites spiral in and collide with their central planet, I wonder whether the Great Red Spot of Jupiter marks the spot where a satellite disappeared into Jupiter? The Great Red Spot is a gigantic atmospheric eddy of turbulent flow. Zeilik (2002, page 202) writes "The Red Spot changes in size, averaging some 14,000 km wide and up to 40,000 km long –it could easily swallow the earth!" The red colour is thought to come from chemical compounds, although the origin of these compounds is not known. That a satellite has disappeared into Jupiter in the relatively recent past is not beyond the realm of possibility. If the Planet Capture Hypothesis is correct, the highly inclined orbit of Triton shows that the capture of this satellite is relatively recent, and it's spiralling in shows that it will come to the same sticky end.

The satellites Io, Europa and Ganymede have eccentric, but prograde orbits around Jupiter. The satellites Pasiphae and Sinope have retrograde orbits around Jupiter.

To conclude this section, I have proposed that planets and satellites have been captured into the solar system. These bodies had varying velocity or speed upon capture, and this led to different forms of orbit; circular, spiralling in or elliptical. Excess energy in a captured body is dissipated in producing an eccentric orbit. An eccentric orbit means that the planet travels a longer distance around the Sun, at a given speed.

Therefore, the motion of planets in orbit around the Sun does not come from acceleration produced by gravity (in-falling); the motion of planets in their orbits comes from the nova or supernova explosion which set the planetary core in motion at its origin.

The motion of planetary cores in their trajectories is maintained by inertia. Inertia is the natural tendency for a body in motion to remain in motion. In my view of things inertia allows planetary cores and meteorite parent bodies to travel through galactic space almost indefinitely until captured.

The gravitational attraction of the Sun on the planets of our solar system has modified the velocity of planets, but has not cancelled the motion that they came with. Consequently, although gravity holds planets in orbit around the Sun, the impulse of the motion of planets comes from outside the solar system.

Speed of orbit

Johannes Kepler (1571-1630) was one of the first astronomers to
describe the solar system and note that planets further away from the
Sun move more slowly in their orbits. Kepler formulated three laws
that describe this motion.

The planets closer to the Sun move more quickly than the planets
further away. Mercury circles the Sun in just 80 Earth days. If Jupiter
travelled at the same speed it would take about 3.5 earth years to
complete an orbit when, in fact, it takes
12 years (Baker 2007, page 12).

Once a planet has been captured into an orbit, the laws of gravity
defined by Newton will cause the planet to conform in its orbital
motion to a certain speed at a certain distance from the Sun. This
produces regularity in the orbits of planets around the Sun. However,
according to my scheme if an incoming planet carries excess energy in
its velocity, this excess energy will be dissipated in a non-circular
orbit. The greater the excess energy, the more elliptical the orbit will
be.

17.2 TABLE OF THE ORBITAL VELOCITY OF PLANETS

Mean orbital velocity

Mercury	47.9 km/second
Venus	35 km/sec
Earth	29.79 km/sec
Mars	24.1 km/sec
Jupiter	13.1 km/sec
Saturn	9.7 km/sec
Uranus	6.81 km/sec
Neptune	5.48 km/sec
Pluto	4.66 km/sec

(The speed or mean orbital velocity of planets in our solar system comes from
the *Encyclopedia Britannica 2011 Standard Edition:* planetary data table).

Spacing of the planets

The radius of each planetary orbit is roughly twice as large as that of the planetary orbit inside it towards the Sun. This is the Titius-Bode rule. It holds as far out as Uranus, but is not true of Neptune, and the orbit of Pluto is a total anomaly since it crosses Neptune's orbit.

The average distance between the centres of the Sun and the Earth is about 150 million kilometres. This distance is called an Astronomical Unit or AU. Mercury orbits the Sun at 0.4 AU, Venus at 0.7 AU, Earth at 1 AU, and Mars at 1.5 AU. The asteroid belt fills in the space of a missing planet between 2 and 4 AU. Jupiter orbits at close to 5 AU, Saturn at 10 AU, Uranus at 20 AU and Neptune at 30 AU.

The spacing of planets has always been thought to be a fundamental feature of the solar system. Taylor points out, however, that there is no correlation between the mass of a planet, composition and spacing. He believes that the spacing of planets is a secondary, rather than a primary property of the solar system. He attributes spacing to tidal forces between planets that rearranged the solar system after their initial formation (Taylor 1998, page 18).

Gamow (1961, page 308) wrote that the spacing between planets is probably dictated by the avoidance of collisions between planets which started with quite elliptical orbits. Any planets which did not obey this rule would have been eliminated by collision. Collision is probably the origin of the ring of debris which is the asteroid belt between Mars and Jupiter. If the orbit of planets were closer together, collisions would have happened because elliptical orbits intercept each other.

I agree with Gamow with respect to the spacing of planets and the avoidance of collisions. According to the hypothesis of capture, a planet entering a clear space will take up an orbit around the Sun that allows it not to collide with any planet-sized body, although it may collide with smaller bodies and survive. The asteroid belt is evidence that one such cataclysmic collision did occur such that debris now orbits in the place of a planet.

The spacing of satellites about the gas giants is also regular, although they are much closer in to the primary planets than the planets are to the Sun. Taylor sees this spacing as a secondary feature that has evolved over time, and he also attributes it to tidal forces (Taylor 1998, page 89). I see it as a secondary feature involving avoidance of collisions. Many collisions may have occurred but they are now testified to only by miscellaneous rubble. The pieces produced by impact will have been sent spinning in all directions and mostly swept up by other planets and satellites.

Tilt

Only one of the planets is rotating in a completely upright position and this is Mercury whose rotation is locked onto the Sun. Jupiter deviates from the upright position by only 3 degrees. The other planets all have significant tilts relative to the plane of the ecliptic. The Earth is tilted at just over 23 degrees and this is the cause of the seasons.

According to the Planet Capture Hypothesis, planets would have been captured with axes pointing in different directions. The tilt of the axis would have been set by the initial conditions that set the body spinning, but tilts can also be modified by the pull of gravity. The gravitational pulls experienced by a planet upon capture could change the angle of the axis of a planet. Planets experience the pull of gravity from the Sun, but also the pull of gravity from other planets in the solar system, especially from the giant planets.

Mars accompanies the Earth with a tilt of 25 degrees. Saturn has a tilt of 26.4 degrees and Neptune has a tilt of just over 29 degrees.

Uranus is a special case; it is lying at right angles to the plane of its orbit with 97.5 degrees relative to the plane of the ecliptic. Because Uranus is lying on its side, its rotation is retrograde. It's five larger and ten smaller satellites orbit the planet in the equatorial plane. So, Uranus represents a whole mini-system whose motion is at odds with the direction of motion of the rest of the solar system. The Nebular Theory has no explanation for this.

Pluto has a tilt of over 119 degrees while Venus has a tilt of 177 degrees which means that these planets are rotating backwards. When the solar system is described in terms of regularity these two examples are again anomalies that defy explanation. On both Pluto and Venus the Sun rises in the west and sets in the east.

Pluto is a very small planet and it is accompanied by a moon one fifth of its size called Charon. Pluto and Charon form a binary system, orbiting a common centre of gravity outside Pluto. They rotate around each other at right angles to the rest of the solar system.

4. The age of planets

Planetary cores shot out of exploding stars have started their journey at different times and in different parts of the galaxy. Some have travelled much further and for a longer time than others. Planetary cores have travelled through different interstellar clouds and accreted mantles turning them into proto-planets. The differences between interstellar clouds have led to the planets having varied compositions in their rocky mantles and atmospheres.

There are two main ways of determining the age of a planet, according to the view I am presenting; the first is the rate of rotation, and the second is the temperature of the core. It is to be expected that a younger planet will have faster rotation and a hotter, molten core. These measures, of course, are relative since the speed of rotation and the heat of the core depend also upon the force of the supernova explosion and the temperature inside the exploding star.

A very old planet is expected to be barely rotating and to have an almost solidified core. However, the age of a planet is quite separate from the time of its entry into the solar system. A planet may be very old, but only recently captured. An old planet may show activity on its surface and turbulence in its atmosphere due to the recent capture event.

Speed of rotation
The speed of rotation of a planet determines the length of a day. It is not proportional to distance from the Sun and it is not proportional to the size of the planet. I am proposing that it is determined by the origin and the age of a planet. Supernova explosions have sent planetary cores spinning on their way; over time the rotation slows down, until eventually it may stop. Speed of rotation is not an absolute measure of age; it is only relative, since each planet had its own initial rate of spin.

Slowing down in the Earth's rotation has been proved by the study of corals. On the Channel 4 programme *Catastrophe: Birth of the Planet* broadcast on 24th November 2008 it was explained that studies of the growth rings in corals that lived 400 million years ago show that at this time a year lasted 410 days and each day lasted 21 hours. Extrapolating this means that 4.5 thousand million years ago, a day must have lasted 6 hours. (I had other information on this subject that gave slightly different figures, but I am using the information given on the programme since it is probably more up to date).

Initial high speed rotation is what one would expect if this momentum was produced by explosion. It is one possible confirmation of the Planet Capture Hypothesis. At the moment of genesis, planetary cores such as the Earth's core may have been spinning so fast that a 'day' lasted only a few minutes or even seconds.

It is thought that the Earth's rotation is slowing down due to tidal friction. This principally involves the influence of the Moon's gravity upon the Earth.

The effect of the Moon's orbit around the Earth is to cause a bulge in the Earth's oceans on the side closest to the Moon due to gravitational attraction and on the side away from the Moon. The bulges move round with the orbit of the Moon. This results in two high tides and two low tides a day, but the height of the tides is strongly influenced by local factors of geography such as the shape of the coastline and the depth of water.

The tidal forces between other planets and their moons, and between the planets themselves would play a role in slowing down the rotation of other planets.

The Earth has a day that lasts 23 hours and 56 minutes. Similarly, Mars has a day that lasts 24 hours and 37 minutes.

The Moon shows us the same face because its rotation matches its revolution around the Earth once every 27 days. This is called synchronous rotation. It results from tidal forces that are effects of gravity between bodies. It is common for satellites or moons to have synchronous rotation. At this point, independent rotation has ended and the motion of the satellite has a lock-on effect.

The slowest rotating planets in the solar system are Pluto, Mercury and Venus. Pluto rotates in a retrograde direction once every 6.4 Earth days. Pluto and Charon are fixed relative to each other, always showing the same face. Mercury rotates every 59 days. It is locked onto the Sun always showing the same face to the Sun due to its proximity.

Venus has such slow rotation that one Venus day lasts longer than one Venus year. Venus has retrograde rotation of 243 Earth days, and its orbit around the Sun takes 225 days.

Jupiter has a rotational period of 9 hours 50 minutes, Saturn 10 hours 14 minutes, Neptune 16 hours 3 minutes and Uranus 17 hours 14 minutes. If these giant gaseous planets are brown dwarf failed stars, rather than planets like the terrestrial planets, then they were not set in motion by supernova explosions. Their rotation cannot be compared to that of the terrestrial planets. They would be rotating because the molecular clouds from which they formed were rotating. It is possible that their rotation does not slow down like that of the terrestrial planets because there are no separate components of solid and liquid to cause tidal friction and allow the transferral of energy in the same way. Do stars maintain their rotation speeds? Is the rotation speed of a star determined by its size? I do not know. The Sun rotates once every 25 days. The rotation speed of the Jovian planets could be compared to that of brown dwarfs if it were known.

Temperature of the core
The temperature of the core of planets in the solar system is not dependent upon their size as one would expect if they had formed at the same time from the same solar nebula by the same processes. I believe that planetary cores started out very hot and have cooled down over time giving a relative indication of their age. According to the view I am presenting, the temperature of the core will depend upon:

- The temperature of the red giant star in which the molten iron core was smelted.
- Size of the core because this affects the rate of cooling.
- Depth of the accreted rocky mantle because this acts like an insulating jacket.
- Time.

It was once thought that the heat of the Earth came from radioactive elements inside the Earth. It is now known that there are not enough radioactive elements present in the Earth's mantle to produce the heat measured.

In fact, the temperature of the Earth's core and the core of other planets is not known for sure. It is thought that the Earth's core is composed of iron and nickel with a temperature that may exceed 6000 K.

An indication of heat in the core of a planet may be the presence of a magnetic field. It is thought that the Earth's magnetic field is produced by a molten liquid core that acts like a dynamo. The other terrestrial planets in the solar system and the moon do not have magnetic fields like the Earth's.

The Moon is thought to have an iron core which is cool and solid surrounded by a rocky mantle. The Moon has an ancient magnetic field which is now dead. Mars is thought to have a molten core, but it has virtually no magnetic field. The magnetic field is 0.002 times the strength of Earth's magnetic field.

Venus is thought to have a molten metallic core, but it has no magnetic field. The solar wind runs right into the upper atmosphere of Venus since there is no magnetic field to act as a buffer.

Mercury does have a magnetic field. Mercury is a very small planet; it is comparable in size to the Moon. However, Mercury has a huge iron core covered by a very thin rocky mantle. Mercury's core represents 42 % of its volume compared to 4 % for the core of the Moon and 16 % for the core of the Earth. Zeilik (2002, pages 187-188) states that it is thought that Mercury has a cold solid core because of its small size, so it was a surprise to find that it has a magnetic field. The magnetic field is only 0.01 times the strength of Earth's magnetic field; nevertheless, it deflects the solar wind in the same way. It may be that Mercury's large metallic core is, in fact, still molten.

Some of the terrestrial satellites have strong magnetic fields, such as Ganymede, a satellite of Jupiter.

The core temperature of Jupiter is thought to be 11 000 K and that of Saturn 10 000 K. They emit heat. However, I have concluded that they are probably old, cool, failed stars, not hot planets. It is not known whether the gaseous planets have cores as the terrestrial planets do. The four Jovian planets all have strong magnetic fields which are thought to be produced by liquid metallic hydrogen that can conduct electric current.

To summarize this section on the age of planets, according to the above criteria the planets can be characterized as follows:

- Neptune, Uranus, Jupiter and Saturn are brown dwarfs of great age.
- Of the terrestrial planets, Mars is similar, but older than the Earth.
- Pluto and Mercury are likely to be older than the Earth.
- Venus is very old indeed.

This leaves the Earth as being the youngest planet, but some of the satellites in the outer solar system may be younger, notably Io (to be discussed later). (Note that the age of the planet is not related to the timing of the capture event).

5. Entry into the solar system

If a planet was captured and took up orbit around the Sun, it would start to sweep its orbit clean. It would accrete all the material that it met in its path. The radiometric dating of rocks on different planets may reveal similar ages because the accreted material from the solar system would be of the same age. According to Gamow (1961, page 306) a planet should take a hundred million years to sweep its orbit clean. This would be true of a planet of whatever origin.

According to the Planet Capture Hypothesis there will be a capture event marked by cratering and volcanism on a terrestrial planet or terrestrial satellite. Upon entering the solar system and taking up a new orbit, a planet would undergo heavy bombardment by impactors of all sizes. The craters left by impactors can often be dated. The period of bombardment would be followed by a period of intense volcanic activity.

Impacts and therefore craters will be found on the leading edge of the planet. The pattern of cratering on a planet's surface will depend upon the axis of rotation of the planet since this defines the leading edge. Extensive cratering on one side of a planet may show that the planet was captured with a pole as the leading edge. A band of cratering would show that the rotation of the planet was producing a revolving edge.

It seems that the edge away from the direction of entry may build up a thicker mantle. This thick mantle may be built up by volcanism leading to the formation of highlands or plateaus. A satellite or moon that had ceased to rotate independently may accumulate mantle more on one side than the other.

Planets that entered the solar system last will not have accreted as much mantle as those that entered early on because there would be less material left to be accreted. Thus, a thin mantle could be indicative of late capture (an example of this is Mercury). This is a corollary of the Planet Capture Hypothesis in which the mantle of planets is viewed as having been accreted as the proto-planet passed through the debris of exploding stars, interstellar dust clouds and the planetary disk of the capturing star with the composition and thickness of the mantle depending on the trajectory.

Planets or satellites covered in thick ice upon entry may not be left with the evidence of impacts. Craters form from impacts to exposed rock surfaces. A thick covering of ice will break the impact and protect the mantle beneath it. One would expect a melting of the ice producing liquid water. (Could the evidence of liquid water on the surface of Mars have been produced by its capture event?)

According to the view of Taylor and no doubt many others, impacts and therefore cratering occurred when the solar system and the planets first formed. According to this view, planets formed at one time about four thousand million years ago; the presence of craters is therefore taken to indicate that a surface is very old. The old surface is said to have preserved features produced in the early solar system when the planet underwent the great bombardment and then undergone little change since. A surface that is classified as young has few craters because it is thought to have been covered with lava from more recent volcanic eruptions. The above classification of surfaces does not apply to the hypothesis I am presenting since the capture

36

events that would have led to cratering and volcanism would have occurred over time, with some events being relatively recent. According to the Planet Capture Hypothesis cratering on the surface of a planet is not indicative of the surface being old.

Craters and volcanoes on terrestrial planets and satellites

Earth: 3.8 thousand million years ago the Earth underwent an intense thermal period of volcanic activity. At this time the Earth's oldest rocks were metamorphosed by heat.

Impact craters are not observed on Earth. This may be due to erosion having erased most of the evidence. If craters once existed, one would expect to find them in areas now covered by ocean. The Earth has a crust on top of the mantle built up over time. The continents are areas of crust that are particularly thick. The crust would have overlaid any craters produced early on in Earth's history.

A few craters have been recognized on Earth's surface. One example is the Sudbury basin in Canada famous as a source of nickel. Nickel would have been brought to Earth by a meteorite. There is the Vredefort structure in South Africa, and Yucatan in Mexico. The Yucatan crater is thought to have been left from the asteroid impact that killed the dinosaurs. These craters have all been produced by random asteroid or meteorite impacts more recently in Earth's history; they do not come from an initial bombardment or capture event.

Moon: The lunar surface is covered in craters of all sizes, some hundreds of kilometres in diameter with concentric rings of mountains. Dark features on the Moon called maria are basins formed by impacts that have filled with basaltic lavas. Maria are only found on the side of the Moon which faces us. The mantle on the near side of the Moon is thinner than on the other side where lavas are more common. The basalt craters have been dated to 3.85 thousand million years old. Thus, impacts only occurred on one side of the Moon and the mantle is thicker on the opposing side.

Mercury: The surface is covered in craters. There are over twenty large multi-ringed basins covered with later impact craters. The surface is covered with dust and rubble like the Moon. Mercury's craters have not been dated since no samples of rock have been obtained.

Mars: The surface is formed of basaltic lava. Mars does not have a crust like the Earth, and there are no crustal features. The southern hemisphere is highly cratered, low terrain. The northern hemisphere has many large volcanoes. Olympus Mons is a huge volcano on Mars that still appears to be outgassing some water vapour that forms thin clouds of ice.

Mars has a large bulge in its mantle on one side of the planet called Tharsis. The bulge is about ten kilometres high at the centre and 8000 kilometres across, covering a quarter of the planet. Volcanic activity has been centred in the Tharsis region for the past two thousand million years, and this has built up the higher area. The Valles Marineris rift valley cuts into the Tharsis region. Thus, Mars illustrates the pattern of cratering on one side of the planet, and a thick mantle with volcanism on the opposing side.

Venus: The southern part has relatively flat rolling terrain with some craters. The northern region is mountainous upland plateaus with no craters. The upland area has thousands of small shield volcanoes, as well as very high volcanoes. Again we see craters associated with low-lying terrain on one side of the planet and volcanoes associated with high plateaus on the other side.

The surface of Venus has the appearance of being relatively young. The 950 impact craters are mostly uneroded. Taylor (1998, page 131) writes, "The craters formed by meteorite impact on the surface of Venus are surprisingly fresh." This state of preservation of impact craters indicates an age for them of between 300 and 700 million years. A few hundred million years ago, Venus' surface was covered with an outpouring of lava, but it has since then only produced a trickle of lava. The present surface of Venus is between 300 and 500 million years old (Taylor 1998, page 132). The steepness of the volcano sides on Venus has been explained by positing that the crust of Venus is exceptionally strong. I suggest that the event which produced the volcanoes may be very recent.

There is another interesting thing about Venus; it has a crust. The northern mountainous region resembles continents on Earth. The low-lying basaltic regions would correspond to the areas of ocean on

Earth, although Venus has no liquid water on its surface. The crust of Venus is folded and faulted like the mountainous regions of Earth, and there are rift valleys. It is thought that Venus may have only one continuous crust that has wrinkled and puckered, rather than crustal plates.

Jupiter's satellites: Jupiter has at least 28 moons or satellites. The four larger ones are the size of small terrestrial planets. They are Ganymede, Callisto, Europa and Io known as the Galilean satellites.

Ganymede: Ganymede shows a cratered terrain. The craters on Ganymede have been dated to one thousand million years old. Ganymede has a metal core surrounded by a rocky mantle covered by an 800 kilometre layer of ice. It has a magnetic field.

Callisto: Callisto is riddled with craters, some filled with ice.

Europa: Europa has a small iron core and a rocky mantle. It is covered with ice and has darker orange-brown areas. Beneath the ice is a salty ocean of liquid water 200 kilometres deep. The surface of Europa is smooth and mostly devoid of impact craters.

Io: Io has a large iron core that generates a magnetic field. It does not have craters, but many volcanoes that have covered the surface with lava as well as sulphur compounds. The volcanoes have built up a rugged, mountainous terrain. What is surprising is that the lava on Io emerges at 1700-2000 K. Magma this hot has not been common on Earth for more than three thousand million years. This, in my opinion, suggests youthfulness in the satellite and a recent capture by the Sun and Jupiter.

To conclude this section, there are indications that the Earth-Moon pair entered the solar system 3850 million years ago. Mars may already have been here. Venus shows many signs of recent capture. The capture of Venus may have occurred only a few hundred million years ago.

6. Small bodies

I am now going to examine the subject of small bodies that enter the solar system. In Chapter 11 I looked at the composition of meteorites and comets; I am now going to look at evidence that they are being captured into the solar system by studying the form of their orbits. I am going to start, however, by looking at the asteroids which come from within the solar system. The section ends with a look at a newly-found beast, the centaur beyond which are found the Trans-Neptunian objects.

Asteroids

The asteroids mainly include 5500 irregular shaped pieces of rock that orbit the Sun in a belt between Mars and Jupiter. There may be another 10 000 smaller pieces. They have varying compositions and are classified as stony, carbonaceous and metallic (the classification is not unlike the classification of meteorites).

Until recently it was thought that the asteroid belt was produced by a massive collision of a large object with a planet. Currently it is the fashion to believe that the asteroids are bits of an unfinished planet that did not manage to bring itself together by accretion. The asteroid belt orbits in the expected place of a missing planet. I believe that the asteroid belt has been produced by collision. The collision would have been produced either by a large body spiralling in to the Sun or by a large body on an elliptical orbit. The collision obliterated a small terrestrial planet that once orbited 2.8 AU from the Sun.

The irregular shapes of asteroids suggest a ring of broken pieces. (One would expect anything shot out of a star in a molten condition to become spherical. This is not the case of asteroids). Smaller asteroids are much more numerous than larger ones; this is a typical pattern for fragmented bodies. Subsequent collisions between asteroids have caused further fragmentation and scattering of asteroids. The evidence for collisions is that asteroids are covered with pits and craters produced by collisions. At one time the asteroid belt may have contained more material than it does now because the orbits of asteroids have led many of them to collide with and be absorbed by surrounding planets, as well as the Sun.

The orbits of asteroids are fairly elliptical, and much more inclined to the plane of the ecliptic than most planetary orbits. The rotation of asteroids is fast; it varies between 2 and 39 hours, with an average of 8 hours. The rapid spinning of asteroids suggests a fairly recent origin or recent collision that set them spinning.

The largest known asteroid is Ceres. It is about a third the size of the Moon. Ceres has an orbital eccentricity of 0.077. This makes its orbit more elliptical than that of the Earth at 0.017, but less elliptical than that of Mars at 0.093. The inclination of Ceres' orbit to the plane of the ecliptic is 10.615 degrees. This is a lesser inclination than that of Pluto at 17.2, but a greater inclination than that of any other planet including Mercury at 7.0 degrees.

There are a multitude of small asteroids of a few kilometres across. The outer asteroids cross the orbit of Jupiter. Some of the nearer asteroids cross the path of the inner planets. The asteroid 3753 orbits between Mercury and Mars crossing Earth's orbit. In fact, a thousand asteroids of over a kilometre in diameter cross the Earth's orbit and if they collided could do deadly damage to the Earth. There are other asteroids that cross the orbit of Mars (Taylor 1998, pages 112-114).

Some asteroids are orbited by other asteroids. The moons of Mars, Phoebos and Deimos, have irregular lumpy shapes and resemble asteroids. Both Taylor (1998, page 61) and Zeilik (2002, page 229) believe the moons of Mars to be captured asteroids.

Meteorites
Meteoritic material is material that falls to Earth captured by the Earth's gravity. Much of it is dust, but meteorites the size of a small pebble up to several kilograms can be found from time to time, especially in Antarctica and in deserts. The largest meteorites that have hit the Earth have been a few metres across.

It has been established that most of the dust which appears as showers of meteors commonly known as shooting stars is debris left behind by the passage of comets. This fine-graded debris burns up in the Earth's upper atmosphere producing a flash of light.

Taylor (1998, page 97) writes that although 99 % of dust grains that burn up in our atmosphere come from within the solar system, some dust grains have velocities so high –greater than 100 km per second – that they must come from outside our solar system. These dust grains are about 20 microns in diameter. This, in my opinion, lends support to the idea that material enters the solar system from outside.

There is also the strong possibility that some meteorites come from asteroid collisions that have shot fragmented pieces across the orbits of other planets.

There is a current idea that some meteorites come from Mars. The organic compounds contained in carbonaceous chondrite meteorites have been taken as evidence that there could be life on Mars. Similarities between the composition of these meteorites and the composition of the rocks and atmosphere on Mars as sampled by the Viking Landers have been taken as support for this view. How the meteoritic rocks would have been launched from the surface of Mars remains a mystery (unless, of course, there really are Martians doing the launching!)

Radioactive dating of carbonaceous chondrite meteorites give dates around 4.6 thousand million years (although some have been dated to 1.3 thousand million years). This is taken to indicate the age of the solar system on the assumption that meteorites are fossils of the so-called solar nebula. I believe that many, though not all meteorites come from outside the solar system.

Zeilik (2002, pages 238-239) writes that iron meteorites show large crystalline patterns if the meteorite is cut and the surface polished. Large crystalline structures such as these only occur when iron cools very slowly from an initial heat of about 1600 K. This could only happen if the iron-nickel meteorite had been part of a larger body at least 100 km in diameter that would slow down the cooling process. These are termed parent bodies. Their subsequent collision and fragmentation is thought to be the source of meteorites.

Zeilik attributes the parent bodies to asteroids formed within the solar system. I propose that parent bodies have been shot out as debris by nova or supernova explosions of red giant stars somewhere in our galaxy. As they travelled through interstellar space, they encountered interstellar clouds containing dust and carbon compounds. The accretion of this material has given meteorites varying compositions. A body composed of a mixture of rocky materials and metals such as iron and nickel may fall apart as it cools down and loses its cohesion. This fragmentation of the parent body has given rise to meteorites with a whole range of smaller sizes.

Evidence for an origin to meteorites outside the solar system is the presence of carborundum or silicon carbide (SiC) grains in the Murchison meteorite, and other meteorites. Silicon carbide is associated with red giant stars and thought to originate in type II supernovas. Microdiamonds are also found in meteorites. These diamonds are thought to form in the outskirts of supernova explosions.

Comets

Comets are conglomerations of dust and ice with a rich mixture of organic molecules. The nucleus of a comet is usually several kilometres across surrounded by a coma of water vapour and other evaporated gases and dust particles. The coma gives rise to a tail of gases and dust.

Observation of Halley's Comet has shown that as the comet approaches the Sun jets of evaporating ice are given off depleting the comet of 20 tons of material per second. Halley's Comet loses a layer a few metres deep of dirty ices with each passage by the Sun. The evaporation of comets shows that they are not ancient occupants of the solar system. They evaporate away and disappear after a certain number of orbits of the Sun.

In Chapter 11, section 12 I proposed that comets are conglomerations of interstellar dust grains. The water ice of comets would come from the mantle of ice which often surrounds interstellar dust grains and the organic molecules would also come from those associated with interstellar dust. I proposed (as did Hoyle and Wickramasinghe 1978) that these 'dirty snowballs' were set in motion by the shock waves of nova and supernova explosions as they move through interstellar dust clouds. The shock wave motion would cause interstellar dust grains to stick together and 'roll' into a large snowball. Kerogen which resembles tar on the surface of some interstellar dust grains would enable the sticking-together process to occur.

Comets typically rotate once every 6 hours. This rapid rotation suggests a recent origin to comets –this means that the shock wave that set them rolling did not occur very long ago, and did not occur very far away. Comets have not travelled very far in galactic terms. Comets travel at high velocities, compared to asteroids and meteorites.

I propose that comets are captured by the Sun in the same way as proto-planets and meteorites from outer space. Comets represent very recent capture events.

There are comets seen on their first visit to the inner solar system, and then never seen again. These comets exit the solar system because they must have enough velocity to reach escape speed from the Sun. Comets that do make regular returns to the Sun are called periodic comets. Comets orbit the Sun in highly elliptical and inclined orbits. Periodic comets are divided into two groups; long-period and short-period comets with the dividing line between the two sets at orbital periods of 200 years.

The long-period comets have highly elongated orbits that are inclined at all angles to the plane of the solar system. Over half of the long-period comets have retrograde orbits around the Sun. The long axes of the orbits are in random directions and the elliptical shapes of the orbits vary a great deal from one comet to another. The eccentric orbits of long-period comets take them way beyond the outermost planets of the solar system. Long-period comets are seen at intervals of over 200 years.

It appears that long-period comets transform themselves into shorter-period comets after a number of transits through the inner solar system. Short-period comets tend to orbit in the same direction as the planets, they conform more to the plane of the solar system and they have much smaller orbital eccentricities. A short-period comet with a period of four years goes out nearly as far as the orbit of Jupiter. One with a period of 40 years goes to the orbit of Uranus and Neptune.

Halley's Comet is a short-period comet with an orbital periodicity of 76 years. Edmund Halley predicted in 1705 that it would be seen in 1758. It was indeed sighted in late 1758 and named in his honour. Halley's Comet was seen during the Norman Conquest of England in 1066 and is shown in the Bayeux Tapestry. The earliest recorded sightings date from 240 BC. The latest sighting of the passage of

Halley's Comet close to the Earth is from November–December 1985 and March–April 1986.

Halley's Comet orbits the Sun in the opposite direction to the planets. It is travelling at 128 000 kph. At aphelion it passes beyond the orbit of Neptune. It has spent around 23 000 years in the inner solar system and has orbited the Sun some 300 times.

It was anticipated that comets would prove to be primitive unaltered samples of the original solar disk. Analysis of Comet Halley has shown that it does not have the composition of the Sun. Iron and magnesium abundances in Halley are the lowest measured on any body in the solar system and do not match with those of primitive meteorites. The primitive nature of comets was misjudged (Taylor 1998, page 97).

Comets provide a test-case of the Planet Capture Hypothesis: It appears to be the case that long-period comets are converted into short-period comets after a certain number of transits through the solar system over thousands of years. I propose that the orbits of comets become less eccentric as they lose energy. The loss of kinetic energy fuelling the trajectory of a comet is progressively compensated for by a more circular orbit. Comets provide examples of recent capture; the gradual conformation of the orbits of comets to the solar system gives insight into the more ancient capture of planets.

Comets start by travelling at high velocity in very elliptical long-period orbits. That the angle of approach to the Sun shows no conformity to the plane of the solar system is an indication of the recent capture. Evaporation with each passage around the Sun at perihelion leaves the comet stripped of matter and exhausted. The gravitational attraction of the planets also affects the course of the comet. Over time the energy of the comet is spent, and the comet becomes a short-period comet. This means that the eccentricity of the orbit has been reduced. A greater number of passages through the solar system leads to greater conformity to the plane of the solar system.

Comets survive only for a short time – measured in tens to hundreds of thousands of years (Taylor 1998, page 92) or less than a million years (Zeilik 2002, page 233). Comets undergo one of several inevitable fates: they dissipate by evaporation and may eventually break into pieces; they collide with a planet; or they are ejected out of the solar system by interaction with the gravity of a giant planet.

Only a few years ago there were excited reports of material about to collide with Jupiter. Comet Shoemaker-Levy was captured into an orbit around Jupiter in 1929. For 65 years it slowly spiralled in until it broke into 25 pieces due to the gravitational pull of Jupiter. Various impacts to Jupiter were reported in July 1994.

The Oort Cloud and Kuiper Belt

The source of comets was not known so it was proposed by a Dutch astronomer in 1950 [3] that there is an unseen cloud of comets surrounding the solar system. This would be a reservoir of comets from which comets plunge towards the inner solar system from time to time. The Oort Cloud, as it became known, would have to contain a hundred thousand million comets to ensure a continued supply. Long-period comets are supposed to come from the Oort Cloud at a distance of 50 000 AU. This left the origin of short-period comets unexplained so a second source was hypothesized, and this is the Kuiper Belt. The Kuiper Belt is supposed to lie in the plane of the solar system at a distance of 100 AU around the solar system although some sources place it at 30 to 1000 AU.

Icy objects have been found orbiting the Sun beyond the orbit of Neptune at 30 AU. These recently-found bodies have been named Trans-Neptunian objects. They will be described below. Some would identify the hypothetical source of short-period comets called the Kuiper Belt with the zone of Trans-Neptunian objects.

The Oort Cloud as a source of long-period comets at 50 000 AU has never been observed and is entirely hypothetical. Comets cannot be seen more than 12 AU from the Sun. The planetary region of the solar system goes out as far as 40 AU. An Oort Cloud at 50 000 AU would make the solar system 1250 times bigger than the orbiting distance of the outermost planet, Pluto. The Oort Cloud is supposed to be formed from material left over as a remnant from the formation of the solar system out of the solar nebula.

---

[3] Oort, J. H. (1950) The Structure of the Cloud of Comets Surrounding the Solar System, and a Hypothesis Concerning its Origin. *Bulletin of the Astronomical Institutes of the Netherlands* Vol.11, No.408, pages 91-110.

It has been suggested by some astronomers that comets come from interstellar clouds. Hoyle and Wickramasinghe (1978) suggested this in *Lifecloud*. Others reject this idea with the argument that no comet has been observed with an orbit consistent with an interstellar origin, and they continue to support the Oort Cloud hypothesis. They argue that no comet has been shown to have a hyperbolic path that would show that it came from outer space. I believe that an understanding of the principles of General Relativity and the curvature of space around massive objects shows that a comet drawn in by the Sun from outer space will assume a curved path. The curved path is determined by the force of gravity and the velocity of the moving object. The distance from the Sun at which curvature started to occur due to the gravity of the Sun and the form of the elliptical orbit could be calculated. Whether this distance coincides with the distance of the hypothetical Oort Cloud or much nearer hypothetical Kuiper Belt could be determined. In any case, I reject any notion of the existence of an Oort Cloud or Kuiper Belt of comets in reserve.

Confirmation of the hypothesis

Comets provide a good test-case for the Planet Capture Hypothesis since their recent capture and transformation from long-period comets into short-period comets indicates the likely mode of transformation which would apply to captured planets.

The recent capture of comets is indicated by their rapid rotation, highly elliptical orbits, non-conformity to the plane of the solar system and high velocity. Halley's Comet travels at 35.5 km/second which is about the same velocity as that of Venus at 35 km/second in the inner solar system.

The orbital motion of comets can be applied as a test-case of the Planet Capture Hypothesis on the basis of the assumption that an object moving at higher velocity will be captured into a more eccentric orbit with the eccentricity of the orbit dissipating the excess kinetic energy. With loss of energy the orbit would become less eccentric and more conformed to a circle although this depends on the mass of the object.

It was after I had conceived this hypothesis and written this chapter that I found information on the orbits of comets in *Encyclopedia*

47

Britannica: Comets: Types of orbits that confirmed the hypothesis. The equation given is as follows: the total energy E of the comet which is a constant of motion, will determine whether the orbit is an ellipse, a parabola or a hyperbola. The total energy E is the sum of the kinetic energy of the comet and of its gravitational potential energy in the gravitational field of the Sun. Per unit mass, it is given by $E = 1/2v^2 - GMr^{-1}$ where v is the comet's velocity and r its distance to the Sun, with M denoting the mass of the Sun and G the gravitational constant. If E is negative, the comet is bound to the Sun and moves in an ellipse. If E is positive, the comet is unbound and moves in a hyperbola. If $E = 0$, the comet is unbound and moves in a parabola.

All periodic comets move in ellipses and have negative energy with regard to the equation. A comet with so much energy that it moves in a hyperbola would not be captured by the Sun's gravitational field. A comet with intermediate energy $(E = 0)$ moving in a parabola would be captured for a short time, but since its velocity equalled escape speed it would pass out of the solar system and not return.

The equation relates kinetic energy to the form of the orbit. A comet with extremely negative energy would presumably be engulfed by the Sun or diverted into hitting a giant planet in the outer solar system or even an inner planet. The equation could be applied to any captured body.

Centaurs

About a dozen small bodies have been found in the outer solar system orbiting the Sun mostly near to or beyond the orbit of Neptune. They are mostly about 200 km in size, and are composed mainly of ice. They are best described as small icy bodies and have been named centaurs. In Greek mythology centaurs were half human and half horse and had a wild and lawless nature. The centaurs have unstable orbits that cross the paths of the giant planets.

A centaur named Chiron orbits the Sun at 10 AU crossing the path of Saturn and swinging out to the orbit of Uranus. It is 175 km across. It appears to be coated in black tar and it gives off gas from time to time like a comet. Its chaotic path is likely to cause it to collide with Saturn or Uranus sooner or later. Pholus is another icy body orbiting between Saturn and beyond Neptune in a highly inclined orbit.

The largest icy body (1996TL66) is 500 km in diameter. It has a wildly eccentric orbit that takes it just beyond Neptune at its closest approach to the Sun and out to 130 AU at the furthest point. The orbit is highly inclined at 24 degrees to the plane of the solar system.

The size of Centaurs of about 200 km (125 miles) in diameter suggests to me that they may have a solid rocky core surrounded by ice. If they were formed of only dust and ice like comets, they would be likely to break apart and disintegrate. Centaurs could be large-sized meteorites which, moving through interstellar dust clouds, became coated in a thick mantle of dust and water ice. The ice of Centaurs would not melt and evaporate away if they remained in the outer solar system where radiation from the Sun is much weaker than in the closer zones, although some evaporation is seen from Chiron in the vicinity of Saturn.

The orbits of Centaurs are more elliptical than those of planets, but less elliptical than those of most comets. The high inclination of their orbits and their instability indicates recent capture. Many Centaurs must have collided with the giant planets; the few that remain are those which have up to now avoided collision.

Centaurs are smaller than Pluto and its companion Charon, but they share many characteristics: their orbits are highly eccentric and inclined; they orbit in the same zone in the outer solar system; and they are icy bodies that probably have a rocky core surrounded by a thick mantle of ice. I would not object to Pluto and Charon being classified in the same category as Centaurs.

Trans-Neptunian objects
The trans-Neptunian objects seem to be an even wilder bunch than the Centaurs. They are found way beyond the orbit of Neptune on the edges of the solar system and include some quite large objects.

The first sighting of a trans-Neptunian object was made in 1992 and since then hundreds have been identified. Quaoar was sighted in 2002 [4] and Sedna in 2004 [5]. Sedna is three-quarters the size of Pluto

---

[4] *Encyclopedia Britannica 2011 Standard Edition:* Year in Review 2002 Mathematics and Physical Sciences.
[5] *Encyc. Britannica 2011 Standard Edition:* Year in Review 2004 Physical Sciences

and has a highly elliptical orbit which takes it from 76 AU to 900 AU distance from the Sun taking 10 000 years to complete its orbit. It was thought that all objects in the outer solar system would be icy and therefore white or grey in appearance, but Sedna is red like the dry, terrestrial planet Mars.

In 2005 a body 1.5 times the size of Pluto was spotted orbiting the Sun at over twice the distance from the Sun as Pluto. It was designated object 2003 UB313, but its unofficial name is Xena. It moves in a highly elliptical orbit that is inclined at about 44 degrees to the plane of the solar system and it takes 560 years to orbit the Sun. Xena is accompanied by a small moon [6].

The orbital non-conformity of captured bodies in the outer solar system is what one would expect if the capture is recent. It is these odd cases that help to confirm the Planet Capture Hypothesis.

Therefore, in the category of small bodies we can define various types:

- Asteroids are the angular pieces of a smashed small terrestrial planet which once orbited the Sun between Mars and Jupiter.

- Some meteorites are rocky pieces shot from nova or supernova explosions. They probably started out as a parent body which underwent fragmentation. Meteorites were captured by the Sun and subsequently captured by the Earth as they passed through the inner solar system.

- Comets are conglomerations of interstellar dust and ice grains rolled together by a shock wave going through an interstellar dust cloud. These fast-moving 'giant snowballs' are captured by the Sun and have limited lifetimes.

- Centaurs are icy bodies that probably have a rocky core —a type of large-sized meteorite that has accreted an icy mantle in its travels and been captured by the Sun into the outer solar system.

---

[6] *Encyc. Britannica 2011 Standard Edition:* Year in Review 2005 Physical Sciences.

- The trans-Neptunian objects are diverse bodies found beyond the furthest planets. The extreme non-conformity of their orbits denotes recent capture.

7. Capture scenario

The following is a scenario of capture events that led to the formation of the solar system as we observe it now taking into account the information that has been given in this chapter:

- The Jovian planets, Jupiter, Saturn, Uranus and Neptune are probably very old brown dwarfs which have cooled down and shrunk in size. They may have been captured by the Sun early on in the life of the solar system.

- Earth, Mars and a planet orbiting where the asteroid belt is located now were captured over 3.8 thousand million years ago. This age corresponds to the oldest crustal rocks on Earth.

The Earth probably captured the Moon before entering the solar system. Craters on the Moon would have been formed by impacts at the time of the capture event of the Earth-Moon pair by the Sun. The youngest ringed basins on the Moon date from 3.85 thousand million years ago.

The major effects on Earth, in this scenario, would have been caused by the capture of planets into the inner solar system passing close to the orbit of the Earth. I propose that planets can cross the path of other planets to take up an orbit closer to the Sun. The degree of inclination of the orbit of a planet to the plane of the solar system (or plane of the ecliptic) is indicative of the order of events since over time a planet's orbit will conform to the plane of the solar system. I propose the following order:

- Capture of Venus some 650 million years ago.

Venus' orbit has an inclination to the plane of the ecliptic of 3.4 degrees. Venus is a very old planet with slow rotation and no magnetic field. Although Venus shows signs of being very old, its volcanism and the turbulence in its atmosphere indicates that it was captured quite recently.

- Capture of Mercury some 245 million years ago.

Mercury's orbit has an inclination of just over 7 degrees to the plane of the ecliptic. Mercury is barely a planetary core with a thin mantle. It retains a magnetic field but has slow rotation with the rotation locked onto the Sun. The presence of a magnetic field shows that Mercury may not be as old as some planets. Its cratered surface is an indication of recent capture.

- A large body captured by the Sun collided with the planet that orbited beyond Mars creating the asteroid belt 66 million years ago. One asteroid sent spinning into the inner solar system collided with the Earth.

The asteroids have highly inclined orbits to the plane of the ecliptic. The asteroid Ceres has an orbital inclination of 10.6. Asteroids sent in all directions will have disappeared into both Jupiter and the Sun, as well as hitting terrestrial planets.

- Capture of Pluto 40 million years ago.

Pluto's orbit has the greatest non-conformity to the plane of the solar system of 17.2 degrees and an odd orbit crossing the orbit of Neptune. This is a sign of recent capture.

The gas giants of the outer solar system may have arrived with some satellites in orbit and captured others since their arrival in the solar system. The following are the satellites that show signs of recent capture:

- Triton has a retrograde orbit around Neptune that is highly inclined to Neptune's plane. Triton is spiralling into Neptune and will one day collide with Neptune sending shock waves throughout the solar system.

- Nereid has a highly inclined orbit around Neptune.

- Miranda has a highly inclined orbit around Uranus.

- Hyperion has a chaotic orbit around Saturn.

- Pasiphae, Sinope, Anake and Carme have retrograde orbits around Jupiter.

- Io which orbits Jupiter has very hot lava pouring from its interior; a sign of youth and recent capture.

In this scenario the order of capture events has been determined by considering the degree of conformity to the plane of the solar system, some other orbital characteristics and characteristics of the planet itself. The actual dates that I have allocated to capture events have been chosen according to the dates of mass extinctions on Earth. The link between extinctions on Earth and events in the solar system is part of another hypothesis that I will present in Chapter 19.

8. Evidence supporting capture

In the last section of this rather long, but important chapter I am going to discuss the evidence that supports the Planet Capture Hypothesis.

Other planetary systems

Planets orbiting other stars cannot be seen by telescopes due to the glare of the star they orbit. However, they can be detected by the gravitational pull of the planet on the star. Jupiter causes the Sun to wobble from side to side about its centre of mass by 12 metres a second. Many stars have now been found to have a wobble. Present detection techniques only allow planets of 400 Earth-masses to be found, while smaller planets escape detection.

It was not anticipated that Jupiter-sized planets would be found orbiting very close to other stars. The first extrasolar planet found orbits the star 51-Pegasi. It has 0.45 Jupiter masses and orbits the star at a distance of 0.05 AU with an orbital period of 4.5 days. A larger planet of 3.9 Jupiter masses orbits the star τ-Bootes with a semi-major axis of 0.046 AU (Shaw 2006, page 208). These orbital distances are much less than the orbital distance of Mercury to the Sun which is 0.4 AU.

That planets resembling the gas giants in size, rather than the small terrestrial planets, should be found orbiting so close to a star contradicts the explanations for planet formation given by the Nebular Theory. The Nebular Theory proposes that planets with a gaseous composition should precipitate out of the solar nebula at the distance of the outer solar system, not within the orbit of Mercury! According to the Planet Capture Hypothesis distance is not a problem since planets of any size could be caught into any orbit, on condition that collision is avoided.

Brown dwarf failed stars can also be captured by larger stars. The newly-found planets may in some cases be old cool brown dwarfs; some have highly eccentric orbits, while others have nearly circular orbits around their captor.

Planetary systems and pulsars

Pulsars are very dense, small-sized neutron stars. A neutron star is the stellar core left behind after a supernova explosion. Many millisecond pulsars are very old, their age being counted in thousands of millions of years.

A millisecond pulsar called PSR 1257 + 12 has more than one wobble in its centre of mass. It is thought to be orbited by three planets. One planet of four Earth masses orbits at 0.36 AU, another of about four Earth masses orbits at 0.47 AU and a third which is much smaller only 0.02 Earth mass orbits at 0.2 AU. These companions all have nearly circular paths (Zeilik 2002, page 378). There are other examples of pulsars orbited by planets.

The question being asked is how would planets around a neutron star survive the supernova explosion which gave rise to the neutron star? With what material would a stellar corpse form planets? Attempts have been made to explain the formation of planets around a dead star from an accretion disk of matter from a stellar companion. But no really satisfactory explanation has been found.

If planets are captured by stars, then a stellar corpse of high mass is just as likely to capture planets as a main sequence star. In fact, a very old stellar corpse is more likely to have planets in orbit around it because it has had more time in which to catch them. The nearly circular orbit of planets around pulsars may be an indication that these systems are very old, since one would expect conformity to a circle to occur over a long period of time. (This phenomenon has been noted in the orbit of comets which start out as long period comets with highly elliptical orbits and transform themselves into short period comets with less eccentricity in their orbits). Therefore, the discovery of planets orbiting neutron stars which are stellar corpses strongly supports the Planet Capture scenario over the Nebular scenario.

Angular momentum in the solar system

It is a curious thing that the plane of the solar system lies tilted at seven degrees to the equator of the Sun. Taylor (1998, page 171) points out that "This is rarely discussed." Could the rotation axis of the Sun have changed since the planetary system was put together? Nobody knows the answer to this problem at the moment. Could it be that captured planets have conformed their orbits to the plane of the orbit of the largest body in the solar system which is Jupiter, rather than to the rotation plane of the Sun?

The distribution of angular momentum in the solar system is a problem for the Nebular Theory (as described in section 1) since the planets are orbiting faster than they should be according to this theory. The angular momentum of the planets is not a problem for the Planet Capture Hypothesis. Each in-coming planet brought its own momentum. The momentum is distributed by the eccentricity of the orbit. Each planet, to survive must obey the spacing rule, and over time it will conform its orbit to the plane of the solar system and to increasing circularity (unless it already had a circular orbit due to entering the solar system with just the right velocity).

Odd cases

There are various odd problems in the solar system —one of these is Uranus lying on its side with its satellites orbiting at right angles to the rest of the solar system. Taylor proposes that Uranus was knocked on its side by a collision, but it would be unlikely that its satellites would rearrange themselves around the equator of the planet in its new position. He suggests that this indicates that the satellites must have formed after Uranus was knocked onto its side (Taylor 1998, page 88). This view is at odds with traditional views of planet formation. Under the scenario I am offering it is likely that Uranus was captured complete with its system of satellites already in orbit (maybe with the exception of Miranda), the complete system coming in at a strange angle.

Spin
Another type of angular momentum –the spin or rotation of a planet also lends support to the hypothesis I am proposing. It is projected back for planet Earth that at its birth it had high speed rotation (this was discussed in section 4). High speed spin would be observed in planetary cores shot out by supernova explosions. The spin would be conserved, but slow down over time. This would explain the different rotation rates of planets and satellites in the solar system. High speed rotation at birth would not be a feature of planets if they precipitated out of a solar nebular, but it is highly suggestive of an explosive origin.

Binary stars
It is now known that binary stars in which two stars orbit a common centre of gravity are very common. What are we to make of binaries in which the stars are stellar corpses?

King and Watson (1986) [7] report on one such case. The report describes two stars orbiting each other every 11 minutes. This implies that the centres of the stars are about 80 000 miles apart (128 720 km). However, our Sun which is a not particularly big main sequence star is 800 000 miles in diameter (1 287 200 km). Only very small-sized, dense stellar corpses such as white dwarfs or neutron stars could orbit each other with their centres only 80 000 miles apart. But how could a binary of dead stars come about? At the end of the main sequence, a star's volume expands massively at the red giant stage, before exploding and collapsing as a stellar corpse. These stars could not have become binaries at their main sequence stage since during the phase of expansion they would have merged together.

The enigma is solved if the binary was formed by capture when the stars had already reached their compact state as stellar corpses.

---

[7] King A.R. & M. G. Watson The Shortest Period Binary Star? *Nature* Vol.323, 4 September 1986, page 105.

Colliding galaxies

The principle of capture appears to apply to galaxies as well as to stars. De Grasse Tyson (1995) [8] writes in a short article that there is compelling evidence that big galaxies have recently captured little galaxies that wandered too close. Evidence includes multiple centres within the same gigantic galaxy. Large galaxies, including the Milky Way and Andromeda, commonly have an assortment of dwarf galaxies in orbit around them. The article reports (De Grasse Tyson 1995, page 69):

"In 1994, R. Ibatha and colleagues at Cambridge University uncovered a set of stars near the galactic center with velocities unlike other stars in their neighbourhood. Subsequent models of galactic capture supported Ibatha's conviction that the group of stars is the partially digested remains of a dwarf galaxy that had been swallowed by the Milky Way within the past billion years. Dubbed the Sagittarius Dwarf, had it been eaten much longer ago, the individual stars would have thoroughly mixed with the surrounding stars and the dwarf would have never been discovered."

Therefore, it seems that large galaxies are capturing smaller galaxies, while binary stars capture each other and this includes stellar corpses. I believe that stars also capture both proto-planets and brown dwarfs to form solar systems (or planetary systems). The same principle applies throughout the universe.
(See footnote [9]).

Features of planets, satellites and moons

I have proposed that the solar system came together with different components. The diverse origin of planets and satellites is supported by:

- Crater dating which is different for each planet and satellite, despite their present proximity.

- The diverse composition of planets.

---

[8] De Grasse Tyson, Neil (1995) When Galaxies Collide. *Natural History* Vol. 12, pages 68-69.
[9] The evidence points to galaxies being able to capture other galaxies. This leads to the possibility that a star such as the Sun could capture a planet from another galaxy. This planet may have unusual characteristics.

- The assortment of rotation rates, angles of axis including retrograde rotation, eccentricity or circularity of orbits, inclination of orbits and both prograde and retrograde orbits.

Therefore, a great diversity of data supports the Planet Capture Hypothesis.

Conclusion

The Planet Capture Hypothesis offers an alternative to the Nebular Theory. I propose that the so-called Solar Nebula of the Nebular Theory has never existed. The view of the solar system based on the observation of general regularity is being swept away by detailed up-to-date knowledge of the diversity of the planets. It is also strongly supported by new observations concerning stars and galaxies.

The Planet Capture Hypothesis explains in terms of capture the inclination of the orbit of planets to the plane of the solar system; elliptical and nearly circular orbits; and differences in the tilt or spin axis of planets. It takes into account the disk shape of the solar system and the spacing of planets.

I propose with this hypothesis that planets are of different ages and origins. This is shown by the speed of rotation and the heat of the core or presence of a magnetic field.

I propose that the planets have been captured at different times testified to by cratering on their surfaces and episodes of volcanism.

The orbits of comets are interesting in that they are a test-case of the Planet Capture Hypothesis. Comets represent recent and frequent capture events. It appears that long-period comets become transformed into short-period comets over time. I propose that the orbits of comets become less eccentric as they lose energy. The orbits of comets give insight into the more ancient capture of planets.

The Planet Capture Hypothesis as a model for the formation of the solar system is fully compatible with Newton's law of gravity and Einstein's understanding of gravity in his Theory of General

Relativity. What the hypothesis has done is simply add a dynamic aspect to the individual elements of the solar system.

A crucial difference between the Planet Capture Hypothesis and current thinking is that current theory views planets as having elliptical orbits because they are falling into the Sun by gravitational attraction (the first case in Newton's thought experiment), while a centrifugal force in the solar system is keeping the planets in orbit. In the Planet Capture Hypothesis the planets are drawn towards the Sun by gravity in what appears to be a centripetal force. The planets trace elliptical orbits around the Sun and remain in orbit because of the excess energy they came with (the second case in Newton's thought experiment).

The case of the satellite Triton spiralling in towards Neptune is a demonstration of what happens to a planet or satellite that is captured with insufficient velocity to resist the force of gravity. The solar system has not been and will not always be the way it is now –the ultimate fate of Triton is to disappear into Neptune. The solar system is a dynamic system and this is only just being realized now.

Bibliography

Abell, George (1969) *Exploration of the Universe* Holt, Rinhart & Winston

Baker, Joanne (2007) *50 Physics Ideas You Really Need to Know* Quercus

Broms, Alan (1961) *Our Emerging Universe* Laurel Editions Dell Publishing Co. USA

Encyclopedia Britannica 2011 Standard Edition: Centaur object; Comets; Gravitation; Halley's Comet; Mercury; Newton, Sir Isaac; Pluto.

Gamow, G. (1947/1961) *One, two, three Infinity* Viking Press

Hoyle, Fred and N.C. Wickramasinghe (1978) *Lifecloud: The Origin of Life in the Universe* J. M. Dent & Sons Ltd

Ince, Martin (2007) *The Rough Guide to the Earth* Rough Guides Penguin Group

Showman, Adam P. & Renu Malhotra (1999) The Galilean Satellites. *Science* Vol.286, pages 77-83.

Solomon, Sean C. et al. (1999) Climate Change as a Regulator of Tectonics on Venus. *Science* Vol.286, pages 87-90.

Taylor, Stuart Ross (1998) *Destiny or Chance: Our Solar System and its Place in the Cosmos* Cambridge University Press

Zeilik, Michael (2002) *Astronomy: The Evolving Universe* 9th edition Cambridge University Press

CHAPTER 18

PLANET FORMATION

Introduction

This chapter is about the structure of the Earth and its formation as a planet. The same principles apply to the formation of other planets as apply to the Earth.

Of the Earth's core and mantle surprisingly little is known, since observation is extraordinarily difficult. The centre of the Earth has not yet yielded its secrets –it remains enigmatic. The explanation I give is an alternative mode of formation for the Earth to that given in current theory, but it is one that accords with Earth's structure.

The origin of Earth's mantle is part of the Planet Capture Hypothesis presented in Chapter 17. In this chapter accretion of the mantle will be discussed in greater detail, and accretion in interstellar space will be put forward as the explanation for the origin of life on Earth.

In this chapter strands of knowledge from astronomy, geology and microbiology will be brought together in an attempt to give new insight.

1. Structure of the Earth

The Earth has a diameter of 12 700 km and a circumference of about 40 000 km. This makes it the largest of the four terrestrial planets of the inner solar system. The Earth has three distinct density layers: the core, the mantle and the crust.

Earth's core

The Earth's core is about the size of the planet Mars. The core extends more than halfway to the surface and is much denser than the rest of the Earth containing almost one third of the Earth's mass. Its density is consistent with it being composed of iron and nickel. Ince (2007, page 120) gives the composition of Earth's core as mostly iron with a few percent of nickel and 10 % of something else – probably oxygen or sulphur since metallic meteorites have this composition.

Earth quakes produce waves that travel through the Earth and can be detected on the other side of the Earth. These shock waves come in two forms: P waves (primary waves) are compression waves which compress rock in the line of travel and S waves (secondary waves) are shear waves which move rock particles up and down. S waves cannot pass through fluids. S and P waves pass through Earth's mantle, but there is a shadow zone with no S waves detected on the opposite side of the Earth to the epicentre of an earthquake since S waves cannot pass through Earth's core. This shows that the core is liquid.

Thus, seismic studies show that the iron core is liquid because it is in a molten state. In the current model of Earth's core the inner core is assumed to be solid because despite being at high temperature, it is under immense pressure. The atmospheric pressure of the Earth's inner core is between 3.3 and 3.6 million atmospheres. The inner core is assumed to be solid because this would explain the dynamo effect (Dynamo Theory will be discussed in Chapter 22).

In the current model of the Earth the core is presented as having a temperature equal to that of the surface layers of the Sun of between 5000 and 6000 K, although the temperature is not known for certain.

That the Earth's large core is liquid is indicated by seismic studies and various other observations. The shape of the Earth is a flattened ellipsoid that bulges at the equator. This is the shape one would expect a massive spinning droplet to have. The rotation of the Earth would cause the fluid of the core to spread out slightly at the equator and be flattened at the poles. The core is encased in the mantle and a thin crust. The two outer layers of the Earth may simply adjust to the shape of the core. Thus, the shape of the Earth shows that the core is liquid.

Another indication that the Earth's core is liquid is the presence of a magnetic field which creates a magnetosphere around the Earth. The magnetosphere protects the Earth from accelerated particles in the solar wind. The Earth's magnetic field is thought to be produced by the liquid core acting as a dynamo.

The mantle

The mantle around the core is 2900 km thick. It is composed of silicate minerals. The upper mantle is composed mainly of olivine and pyroxene. Olivine is typically an olive-green colour. It is a magnesium iron silicate with the general formula $(Mg,Fe)_2SiO_4$. Olivine consists of isolated silicon-oxygen tetrahedra (SiO_4) joined by ionic bonds. Pyroxene is also greenish in colour and the silicate crystals commonly contain calcium, iron and magnesium, and sometimes sodium. In pyroxene the silicon-oxygen tetrahedra form chains with the formula $(SiO_3)_n$

The material which erupts from volcanoes comes from the mantle. It is called volcanic basalt and is composed of olivine, pyroxene, amphiboles and feldspar. Amphiboles and feldspars are aluminosilicates which means that they contain aluminium as Al_2O_3. Amphiboles have double chain SiO_4 tetrahedra that generally contain iron and magnesium, as well as hydroxyls (OH) and halogens (F, Cl). Amphiboles are found in granite. Feldspars are framework silicates which incorporate potassium (K), sodium (Na) or calcium (Ca), and rarely barium (Ba).

The mantle can be divided into several zones; the lower mantle has denser silicate minerals compressed by the overlying layers. The lower mantle is thought to reach a temperature of 4000 K. The upper mantle is less compressed and is thought to reach a temperature of 1700 K. The uppermost layer of the mantle is known as the lithosphere. The lithosphere is the cooler outer part of the mantle. It has a maximum thickness of 120 km beneath deep ocean floor.

When earthquakes occur they release vibrations that pass through the Earth and make it ring. As already mentioned, earthquakes set two types of vibration in motion, primary waves (P waves) and secondary waves (S waves). The S waves show that the Earth's mantle and crust are solid. Lamb & Sington write (1998, page 98) "the mantle must have the strength of steel to explain the speed at which the S waves travel." I take the seismic indications that the Earth's mantle consists of solid rock at face value. (In Chapter 21 the hypothesis that there are convection currents in the mantle causing the rock to rise, move sideways at the surface and descend again will be discussed under Plate Tectonics).

The mantle in general is not molten; however, pockets within the mantle become molten on occasions for reasons that are not entirely understood. The pools of molten rock form magma chambers that eventually burst to the surface as volcanic eruptions showering molten rock upon the surface of the Earth (the reasons for volcanism will be discussed in Chapter 19). The rock that emerges from volcanoes as lava or magma solidifies as basalt. The composition of basalt reveals the composition of the mantle.

Earth's crust
Earth is special in that it has a crust. The Earth's crust is on average 7 km thick beneath the oceans, although it is surprisingly hilly beneath the sea. The continental crust extends on average 35 km beneath continents lying on top of the denser mantle. Mountains on land often rise several kilometres high. The highest mountain, Mount Everest, stands at 8848 metres (or 8.8 km) above sea level. Underwater mountain chains may be 2500 metres high.

The surface temperature of the Earth is 288 K or 15 °C. If one tunnels into the Earth's crust, the deeper one goes, the hotter it is. The deepest mines such as some gold mines in South Africa penetrate 5 km into the ground, and must be artificially cooled by ice-chilled air for miners to work there. The temperature of the crust increases by 10-15 °C for every kilometre down (Lamb & Sington 1998, page 104). One would expect the temperature of the crust resting on the mantle below continents to be at 350-525 °C.

The crust of ocean basins and the lowest layers below continents are basalts. Basalt is ubiquitous among the planets of the solar system. Basalts come from magma that has reached the surface of the Earth through volcanic activity. It originates from melting of the mantle.

The continents are made up of granite rather than basalt. Granite is formed as molten rock underground. It is later exposed to the surface by erosion as granite intrusions into sedimentary rock strata.

Much of the Earth's crust is made from sedimentary rocks. The sedimentary layer on top of the basalt rock of ocean basins is on average 1 km thick, with thicker sediments around landmasses. The sedimentary rock strata of continents are much thicker.

Continental crust appears to be unique to the Earth. It is composed mainly from sedimentary rock strata and granite intrusions. The sedimentary rocks of continents are often formed from deposits of the tiny shells of deceased forms of life. The calcium carbonate of limestone was originally deposited by calcareous algae and corals forming reefs in shallow seas, and by marine foraminiferans. Sediments accumulated in marine and freshwater environments later became uplifted and turned into sedimentary rock strata.

Volcanoes blow out clouds of dust, ash and rock debris with particle sizes ranging from the consistency of flour to sand particles to larger pieces of rock. This loose material can also be consolidated into sedimentary rocks. Some sedimentary rocks are formed from grains that have been weathered out of previous rocks and redeposited by wind and water.

Sedimentary rocks and basalts can be transformed into more resistant metamorphic rocks by heat and pressure. Metamorphic rocks do not experience erosion to the same degree as softer rocks.

Radioactive dating shows that the oldest rocks found belonging to Earth's crust are 3.8 thousand million years old. I take this as the date of the earliest formation of Earth's crust. The current model places the Earth as 4.6 thousand million years old, with the oldest rocks missing from the record.

Hydrosphere
The Earth has an important hydrosphere of liquid and frozen water forming oceans, lakes, rivers, glaciers and groundwater (groundwater resides in subsurface aquifers). The total mass of water in the oceans equals about 50 % of the mass of sedimentary rocks now in existence and about 5 % of the mass of the Earth's crust as a whole (*Encyclopedia Britannica 2011 Standard Edition*: Hydrosphere).

Oceans cover nearly 71 % of the Earth's surface with an average depth of 3795 metres. Exposed land occupies the remaining 29 % of Earth's surface and has a mean elevation of only 840 metres (*Encyclopedia Britannica 2011 Standard Edition*: Ocean). By volume, 97.96 % of the water on Earth exists as oceanic water and associated sea ice. Glaciers and ice caps hold 1.64 % of the Earth's water – this may have risen to 3 % during the height of glaciations during the Pleistocene. Groundwater represents 0.37 % which is ten times more abundant than the water contained in lakes and streams at 0.036 %. The atmosphere contains a mere 0.001 % of Earth's water in gaseous phase or as droplets (*Encyclopedia Britannica 2011 Standard Edition*: Ocean).

Atmosphere
The surface of the Earth is surrounded by an atmosphere of gases up to 1000 km altitude where molecules escape into space.

Earth's atmosphere is composed mostly of nitrogen and oxygen. Carbon dioxide currently represents only 0.03 % of the atmosphere. In former times the atmosphere of the Earth had much higher levels of carbon dioxide and the planet was generally warmer than today. In the Cretaceous Period some 100 million years ago, the Earth had about three times more carbon dioxide in the atmosphere than it does now. Much of this atmospheric carbon dioxide has been locked up by the formation of carbonate sedimentary rocks such as limestone.

2. Accretion of the mantle

The classic theory for the formation of the Earth goes along the following lines:

The solar nebula expanded and cooled down until droplets of iron and particles of mineral silicates condensed out of the gas. This dust coalesced into lumps of material orbiting the Sun. These small boulder-sized objects collided and stuck together to form planetesimals. The Earth grew by the accretion of planetesimals whose impacts left craters on its surface.

At this stage the Earth was an amorphous mass with no differentiation into core, mantle and crust. Heat produced by impacts caused the whole body to melt and differentiate. When this happened the heavier elements such as iron sank to the centre of the planet to form the core, while the lighter silicates floated on the surface and solidified to become the mantle. When the surface was sufficiently cool oceans condensed from water vapour and continents appeared.

I do not think that this is an adequate explanation of the distinct features of Earth's structure. The core, the mantle and the crust are discrete entities with defined boundaries. I believe that each of these features has its own separate origin.

In Chapter 17 I wrote about planetary cores as part of the Planet Capture Hypothesis. I proposed that finely divided iron and silicate dust are drawn into red giant stars. Inside the star the iron is smelted into globules of molten iron. When a nova or supernova explosion occurs, the globules of iron are shot out and begin to travel through space. These globules of molten iron and nickel become the cores of planets.

With the Planet Capture Hypothesis it is proposed that these planetary cores of molten iron move randomly through interstellar dust clouds as they travel through the galaxy. As they move through the dust clouds the planetary cores accrete cold silicate dust, water ice and organic carbon molecules by gravitational attraction. The silicates would build up as a layer forming mantles around molten iron cores.

The heat of the core may cause many carbon compounds to be converted into gases such as carbon dioxide and carbon monoxide. The water ice would be vaporized. These gases would start to form a very thin atmosphere around the proto-planet. Silicates, on the other hand, can withstand heat and would remain as a jacket around the molten core. Silicates when heated remain as silicates and so form rocks, while organic molecules when heated become carbon dioxide gas and other gases and so form an atmosphere of gases.

Accretion turns the body into a proto-planet composed of a core and a mantle with the beginnings of an atmosphere.

One of the principle minerals of the Earth's mantle is olivine. Olivine has been identified in the surface rocks of the Moon and of Mars, and so it appears to be common among planets and satellites. There are three lines of evidence that support the hypothesis of the formation of Earth's mantle from interstellar dust.

The first evidence is the discovery of magnesium-rich olivine as a component of carbonaceous chondrite meteorites that also contain iron and nickel. The second evidence is that the spectral signature of olivine has been seen in the dust disks around young stars. The third evidence is that the spectral signature of olivine has been identified in the tails of comets (see footnote [10]). The discovery of olivine suggests a link between these locations in space and the mantle rocks of the Earth.

The interstellar dust clouds which a planetary core passes through may be diverse in nature. There may be clouds of debris surrounding the exploded star containing material with radioactive elements cooked up inside the star before the explosion. There may be zones with interstellar dust grains associated with carbon-based life and water. Other clouds may contain dissociated molecules where complex molecules have been broken down by ultraviolet radiation. Other clouds may contain finely divided silicates and iron and no life. The composition of different interstellar clouds will affect the composition of the accreted mantle.

Accretion turns a planetary core into a proto-planet. If the proto-planet approaches a star, it may be captured. As the proto-planet that became the Earth entered the solar system, it would have accreted debris found in the vicinity of the Sun and swept its orbit clean. When a star captures a certain number of planets, they sweep away the cloud surrounding the star (by accretion onto the planet) and this allows the star to shine and be detected by telescopes as a source of light.

Therefore, there would be distinct phases in accretion of the mantle; there would be an accretion phase in the vicinity of the exploded star, the passage through interstellar dust clouds of various types, and finally an accretion phase in the solar system disk of destination.

The mantle of a planet would be accreted as cold material. Initially the accreted material would be heated to a high temperature by the

---

[10] Campins, Humberto & Eileen V. Ryan (1989) The Identification of Crystalline Olivine in Cometary Silicates. *The Astrophysical Journal* Vol. 341, pages 1059-1066.

molten core, but as the mantle increased in thickness, its surface in contact with the coldness of outer space would become much cooler. The temperature gradient from the molten core to the coldness of outer space would mean that when a certain thickness of mantle was reached, the surface of the mantle would be warm, not hot. At this stage the proto-planet could accrete microbial life during its passage through interstellar clouds containing life.

Chemolithotrophic Archaea convert carbon dioxide and hydrogen into organic cell material, methane and water, even in total darkness. Accretion of Archaea could lead to accumulations of organic material on the surface of the proto-planet which would become sandwiched-in when the next phase of accretion occurred. Pockets of trapped organic matter may be transformed by heat and pressure. Organic matter composed principally of carbon may be transformed into diamonds. This is the subject of section 6 of this chapter.

Thus, I propose that the Earth's mantle formed by the accretion of debris from exploded stars, interstellar dust and matter found within the solar system. It originated as cold material, but became heated by proximity to the molten core. The molten core acquired its heat from within the star of its origin.

3. Age of the Earth and cooling

With a new model to explain how the Earth formed it is profitable to revisit calculations concerning the rate of cooling of the Earth and the age of the Earth.

In the 18th and 19th centuries it was assumed that the Earth had been formed in a molten state and had gradually cooled down; the outer crust had become solid, while the interior of the Earth remained molten. By considering the temperature of the Earth now and measuring the rate of cooling of various solids, estimates of the age of the Earth were attempted. A Frenchman called Georges-Louis LeClerc, Comte de Bouffon, (1701-1788) heated up balls of iron of different sizes and measured their rate of cooling; this gave him an estimate of the Earth's age of 75 000 years.

69

Lord Kelvin

William Thomson, also known as Baron Kelvin of Largs or Lord Kelvin (1824-1907), an Irish physicist –also attempted a geophysical determination of the age of the Earth. Lord Kelvin was impressed by the fact that the deeper miners go down into coal mines, the hotter the surrounding rocks. This shows that heat is flowing out of the Earth. Kelvin postulated that at its formation, the Earth had an initial temperature throughout its mass the same as that of magma which now erupts from volcanoes. The temperature of molten basalt is 1100 °C. He measured the thermal properties of various rocks to see how long they take to cool down. He calculated how long it would take a body the size of the Earth to cool down such that the temperature gradient would match that observed in coal mines. In 1862 Thomson calculated the age of the Earth to be up to 400 million years old. Later he revised the age to between 40 and 20 million years. (See footnote[11]).

Calculations of the age of the Earth based on heat loss may have been wrong, but they showed that the Earth which previously had been supposed to be of infinite age was, in fact, of finite age. This encouraged geologists to seek other ways of calculating the Earth's age such as by the rate of accumulation of sedimentary strata. By the turn of the century the consensus of opinion among geologists was that the Earth was 100 million years old.

Rutherford

Radioactivity was given its name by Marie Curie; however, it was Ernest Rutherford a New Zealand-born British physicist working with Frederick Soddy who formulated an understanding of radioactivity in 1902. They proposed correctly that radioactive elements are unstable and spontaneously decay into a more stable form which may be another element. To do this the elements emit radiation and heat. Rutherford realized the implications of the presence of radioactive elements within rocks for the calculations proposed by Kelvin concerning the age of the Earth. If heat was constantly being

---

[11] As a side note Thomson or Lord Kelvin opposed the views of Charles Darwin and this led to his unpopularity amongst Darwin's supporters, notably Thomas Huxley. This seems to have over-shadowed Kelvin's many scientific achievements in engineering and physics in the way that he is remembered today, although he was highly acclaimed during his lifetime.

produced by radioactive elements within the Earth, then the heat measured did not come from the Earth's formation and if this is the case, Kelvin's calculation of the age of the Earth was totally wrong.

The view that much of the Earth's heat comes from radioactive decay prevailed for well over half a century. Friedlander writes in 1956 (page 393) much of the Earth's uranium is concentrated in the crust. The crust of the Earth contains enough uranium, thorium and potassium to supply more than half of the average heat loss of the Earth, about 10^{-6} calorie cm^{-2} sec^{-1}. However, more recent research has shown that radioactive elements do not supply this amount of heat.

The age of the Earth issue has become closer linked to another aspect of radioactivity – in 1905 Rutherford proposed that the radioactive decay of certain elements could serve as a clock. When rock is molten the clock is set to zero. After cooling it takes about 700 million years for half the atoms of uranium-235 to decay to lead-207. This is the half life of uranium. The proportion of uranium and lead in a sample of granite can thus be used to calculate the length of time since the granite emerged as molten magma.

The method proposed by Rutherford became radiometric dating used today to date rocks and fossils. Radiometric dating showed the oldest rocks on Earth to be 3.8 thousand million years old, and therefore the Earth must be somewhat older than this. This is a separate line of argument from the heat loss argument.

Current theory on the source of Earth's internal heat
The current generally accepted theory is that the Earth is 4.6 thousand million years old and all bodies in the solar system are deemed to have the same age and to have formed at the same time. The heat inside the Earth is now thought to have originated during its initial phase of formation by the accretion of planetesimals. The impact of each planetesimal would produce kinetic energy heating up the Earth as a body. This internal heat of formation is still flowing out. It has been found that heat generated by radioactive decay provides some internal heat, but does not, in fact, change by very much the time taken for the Earth to cool down.

The Earth's core is thought to have a temperature of 5000 to

6000 °C. This temperature is calculated by the assumption that the inner core composed of iron and nickel is solid. The interface between the outer and inner core is assumed to be the melting point of iron and nickel. The high pressure of the core means that this temperature is much higher than on the surface of the Earth and experiment shows this temperature to be 5000 °C (Lamb & Sington 1998, page 101).

Within the Earth's crust the temperature increases by 10-15 °C for every kilometre down. Lamb & Sington (1998, page 104) point out that if the temperature continued to increase with depth at a constant rate, the centre of the Earth would be at 80 000 °C, not 5000 °C.

According to the current model this temperature gradient does not apply because the surface of the Earth's mantle is losing heat quicker than the lower layers. The reason for this, according to the Theory of Plate Tectonics, is that the Earth's mantle is losing heat by convection, and not by simple conduction. The convection model requires that the minerals in the Earth's mantle rise to the surface, lose heat at the surface and drop down again in circular motions. This, they claim would make the temperature gradient deep within the Earth less than near the surface, as measured in the Earth's crust.

Alternative hypothesis concerning the Earth's internal heat
In the model that I am proposing, the different layers of the Earth are seen as quite distinct in origin as well as in composition. The core is molten and is the source of the Earth's internal heat. If the core was shot out of a red giant star by nova or supernova explosion, its initial temperature would be determined by the temperature of the star. The size of the star would be important as would the depth within the star at which the globule of molten iron and nickel was formed since the surface temperature of a star is very different from its core temperature.

The mantle is not a contributor to internal heat, but an insulating jacket around the core. The mantle would have been accreted as cold matter, and have gained heat from the hot core. The mantle is under extremely high pressure and is solid, except in pockets that melt and then emerge at the surface as magma. I believe that the mantle is conveying heat from the core to the outside by conduction. The conductivity of the mantle and the temperature gradient within it will be quite different from that of the core. I reject the notion that

72

movements are taking place in the Earth's mantle causing it to lose heat by convection currents. The crust is another thin outer jacket that insulates the core. The mantle and crust slow down heat loss from the core.

Returning to Lord Kelvin's calculation of heat loss and the age of the Earth, one problem is that he assumed the whole Earth as a body to have started out at the same temperature and the second problem is that the temperature assumed to apply is far too low. It is believed today that the Earth's core is far hotter than the 1100 °C proposed by Kelvin.

I believe that calculations of the rate of heat loss from the core to the outside would make it possible to estimate the age of the Earth using the model I am proposing; however, the temperature of the Earth's core now is not known, and the temperature of the star which gave birth to the planetary core is not known. If data on one of these became available in the future, the calculation could be done. The time that the Earth takes to cool down will be lengthened by the presence of radioactive elements producing heat, but this is only something to be added into the calculation.

The rate at which heat flows out of the Earth's interior to the surface and then into space is known. The total outward energy flow averages 0.06 watt per square metre at the surface. Over the whole surface area of the Earth the amount flowing out per second is large –about 10^{13} joules per second (Zeilik 2002, page 158). (See footnote [12]).

Taking the temperature gradient inside the Earth's crust of a 10-15 °C increase in temperature for every kilometre down and multiplying it by the thickness of the crust and mantle, the following calculation indicates what the temperature of the Earth's core could be:

Thickness of mantle = 2900 km Average width of crust = 35 km
Surface temperature of Earth = 15 °C

$2900 + 35 \times 10 = 29\,350 + 15 = 29\,365$ °C
$2900 + 35 \times 15 = 44\,025 + 15 = 44\,040$ °C

---

[12] The Earth continuously loses heat, but the surface temperature is more influenced by the energy added by sunlight to the air, ground and oceans. The Sun's flux at the Earth is 1370 watts per square metre.

It seems a bit daring to suggest that the Earth's core could have a temperature of between 29 600 and 44 300 K (see footnote [13]). This would make the Earth's core as hot as what is thought to be the temperature of Jupiter's core.

Comment

The age of the Earth is no longer an issue that needs to be addressed if radiometric dating is reliable, however, an examination of the internal heat of the Earth would allow one to choose between different models describing the formation of the Earth and other planets.

I do not believe that the planet formed by the gravitational accretion of planetesimals. I, therefore, do not believe that the internal heat of the Earth came from this source. I do not deny, however, that impacts from meteorites and asteroids would generate heat in the localities of the strikes.

One of the notions applied to planets is that of evolutionary lifetime. The evolutionary lifetime of a planet depends on its store of internal heat and the rate of loss of heat to space. When a planet has lost its heat and become cold, its evolutionary lifetime has expired. The evolutionary lifetime of a planet is said to depend entirely on its size. The larger a planet, the longer it's evolutionary lifetime.

"A planet's total internal energy basically depends on its volume. The rate at which a planet radiates energy into space depends on its surface area. If you divide amount stored (volume) by the loss rate (surface area), you get the evolutionary lifetime (which depends directly, then, on the radius)." (Zeilik 2002, page 189).

If it were found that the temperatures of planet cores are not related directly to volume, and there are indications that this is the case; then the model that I am proposing could be examined as an alternative.

---

[13] To convert centigrade ($^{\circ}$C) to Kelvin (K) add 273. Zero degrees Celsius or centigrade is the freezing point of water and 100 $^{\circ}$C is the boiling point of water on Earth's surface at sea level. Zero Kelvin is absolute zero (-273°C) which is the temperature at which all motion of molecules ceases.

4. Accretion of unicellular life

I have presented the notion that the mantle of the Earth (and other planets) was accreted as cold material during the passage of the proto-planet through space. The Earth's mantle is heated by the molten core; however, the uppermost layer of the mantle called the lithosphere is relatively cool. At a time in the planet's history when it did not have a crust covering the mantle, the exposed surface of the mantle is likely to have been cool enough to support life, while not being too cold due to the stored energy within the proto-planet.

These conditions would make it possible that viable interstellar unicellular life could be accreted along with the accretion of interstellar silicate dust – if there is such a thing as interstellar unicellular life. We saw in Chapter 11 that interstellar unicellular life, if it exists, would be closely associated with dust grains.

Seeds of life

It has been discussed in the context of theories relating to life being seeded upon the surface of the Earth from space such as proposed in the Theory of Panspermia, that no form of unicellular life could withstand the passage through the Earth's atmosphere. The Earth's atmosphere would cause any incoming seeds of life to burn up, as meteors do.

This problem would not occur when the Earth was at the stage of being a proto-planet since at this stage it would not have a significant atmosphere. It would only have the gases associated with the recycling processes of life close to its surface. Thus, accreted material would not have to pass through a thick atmosphere when accreting onto a proto-planet.

Another objection to the notion that the Earth was seeded with pre-existing life is that life is unlikely to survive travelling through space unshielded from ultraviolet radiation. To this I would reply that interstellar life, if it exists, is shielded by its association with interstellar dust. The dust of dark clouds causes light to bounce around inside dark clouds lengthening the wave length and reducing the harmfulness of strong radiation. The water and organic molecules associated with dust grains may, in fact, constitute a gel secreted by a life form that also protects it against harmful radiation. Life, if it was

75

picked up at all by the Earth, was picked up *in situ* within a dark cloud. Life itself did not travel through space; it remained within the cloud until picked up by a travelling body such as a proto-planet or meteorite.

I, therefore, believe that some of the objections to the Earth being seeded by extraterrestrial life can be answered. I envisage that unicellular life could have taken the Earth through various stages that I will now explain.

Stages in the first colonization of proto-planet Earth

I am going to present a scenario in which the Earth becomes inhabited by unicellular life before it came into orbit around the Sun:

1. As the proto-planet Earth travelled through the darkness of outer space, before capture into the solar system, its lithosphere surface of basalt rocks could have been inhabited by anaerobic chemolithotrophs accreted to the surface from dark clouds. This stage represents life before light.

Methanogens live on hydrogen gas and carbon dioxide or organic matter, producing methane and water. They can live in total darkness because they obtain energy from chemical processes rather than from light. The stored heat from Earth's core could have made this possible by offsetting the cold conditions of outer space. These methanogens may have lived beneath the surface in spaces in between accreted material or beneath a covering of ice. Methanogens continue this habit today living in subsurface communities of bacteria inside rocks and inside glaciers.

The rocks of the Earth's crust were once thought to be essentially sterile; however, there are indications now that there may be a vast biolithosphere yet to be explored. Prescott (1999, pages 882 and 886) writes that viable microorganisms have been found to a depth of 1000 to 1200 metres below the surface. Subsurface bacteria are found living beneath the sea floor, in deep continental oil reservoirs and in rocky outcrops.

Thus, during the outer space phase of Earth's journey, the proto-planet may have been inhabited by anaerobic bacteria living in

76

darkness upon the gases hydrogen and carbon dioxide, organic matter and also upon the oxidized compounds of basalt rock. During this phase the reduced gases methane, ammonia and hydrogen sulphide would be produced; also, abundant water and reduced metals such as ferrous iron (Fe II) which would become dissolved in the water.

2. When Earth approached the Sun and took up orbit around the Sun a new phase might have dawned –that of anaerobic bacteria practicing anoxygenic photosynthesis using light from the Sun.

Green and Purple sulphur and non-sulphur bacteria practice anoxygenic photosynthesis and use carbon dioxide and either hydrogen sulphide or hydrogen to produce organic matter and water, and they often deposit elemental sulphur. Under these conditions Earth would again see an accumulation of water.

3. Organisms that practice aerobic photosynthesis appeared when there was sufficient water accumulated upon the surface of the Earth.

Cyanobacteria practice aerobic photosynthesis which means that they use the energy from sunlight to split water releasing oxygen and building up organic matter.

The first indications of free oxygen in Earth's atmosphere are the banded iron formations. The earliest banded iron formations date from 3.8 thousand million years, but they are more common by 3 to 2 thousand million years ago. Banded iron formations were formed by the seasonal release of free oxygen into water in a generally anaerobic atmosphere. This converted dissolved ferrous iron in the water into insoluble ferric iron (Fe III) with the ferric oxides being laid down in layers. The layers correspond to the photosynthetic activity of summertime releasing oxygen, sandwiched between the die-off of winter.

Photosynthetic algae joined cyanobacteria in producing free oxygen by aerobic photosynthesis, and at a later stage plants continued the production of oxygen. When oxygen in the atmosphere reached a certain level aerobic respiration became possible and the Earth became colonized first by protists and later by multicellular animals.

Aerobic respiration depends on free oxygen and returns water and carbon dioxide to the environment.

The aerobic phase of the Earth is marked by a cycling of water and free oxygen, whereas the initial phases saw a net production of water and the beginnings of an atmosphere. These three stages show how Earth as a proto-planet could have been inhabited by microbial life whose adaptation and colonization represented the first steps in transforming the Earth from a dry, sterile lump of rock into a fertile world.

Six stages in the biological transformation of planet Earth are described in Chapter 10, section 10 of Part I. The above stages delineate the first three stages in this scenario. These three early stages of planet Earth are depicted on the covers of Parts I, II and III of *The Steps of Creation* with a caption on page vii.

5. Diamonds

The brilliance of a diamond comes from its purity and regular crystal structure. The value of a diamond depends on its brilliance. Diamonds may also prove valuable to the notion of mantle accretion which is part of the Planet Capture Hypothesis since the presence of diamonds in the Earth's mantle appears to confirm the idea that firstly, carbon material could have been accreted onto the mantle surface in space and secondly that life may have inhabited an ancient surface accumulating carbon material in layers and pockets now buried.

Diamonds are formed within the Earth's mantle at a depth of 150 to 300 km (i.e. 115 – 265 km below the Earth's crust). Volcanic eruptions sometimes bring diamonds to the surface sprinkled in amongst basalt rock of olivine and pyroxene.

Diamond is the only gemstone composed of a single element –carbon. The carbon atoms form a tightly-packed, strongly bonded crystal structure which gives diamond its qualities of hardness and light reflection. Kerogen composed of hydrocarbons (carbon and hydrogen atoms) or coal may be transformed into graphite. Graphite is a grey-coloured, layered compound consisting of hexagonal lattices of only carbon atoms. Graphite is transformed into diamond in the Earth's mantle by high pressure and high temperatures of between 900 and 1300 °C which causes the carbon atoms to assume a three-dimensional structure.

The diamonds that come to the Earth's surface are generally very old, ranging from a thousand million to 3.3 thousand million years old. Diamonds are found in the volcanic pipes of ancient volcanoes in the oldest areas of continents. These areas are often characterized by ancient basalts metamorphosed into serpentine rock. Diamonds have often been eroded out of the volcanic pipes and are now widely dispersed in alluvial deposits.

The ratio of ^{13}C and ^{12}C isotopes shows that some diamonds from the Earth's mantle are of inorganic origin, while others are of organic origin. This has been rationalized by assuming that diamonds of organic origin come from the carbon material of life living on the surface of the Earth that became pushed beneath the Earth's crust by subduction (subduction occurs in ocean trenches, not generally on land, however).

I suggest that the carbon material that forms diamonds found its way into the Earth's mantle by an extraterrestrial route. As I have already suggested, the Earth may once have had a surface which was the basalt surface of the mantle. This ancient surface could have been inhabited by chemolithotrophic bacteria before the Earth entered the solar system. There would have been a build-up of organic material on this surface. When the Earth entered the solar system, it accreted a new layer of mantle sealing in the layer of living matter. These buried pockets or layers of organic kerogen material could have been converted into diamonds by the weight of the mineral layers on top and heat emanating from the Earth's core.

Synthetic diamonds are made by subjecting graphite and a metal catalyst to high pressure (7 gigapascals) and high temperature of over 1700 °C. It is interesting that diamonds can also be made from methane gas. Laura Benedetti et al. (1999) [14] have shown by experiment that methane can be induced to form diamond at pressures of 10-50 gigapascals and temperatures of 2000-

---

[14] Benedetti, Laura Robin et al. (1999) Dissociation of CH_4 at High Pressures and Temperatures: Diamond Formation in Giant Planet Interiors? *Science* Vol. 286, pages 100-102.

3000 K. The high temperature breaks the C–H bonds by pyrolysis and the high pressure condenses the carbon. Doubly and triply bonded hydrocarbons form in the cooler areas, and are the precursors to diamond.

In 1961 shock-wave methods and explosive-shock techniques were first used to produce diamond powder. Beginning in the 1950s Russian researchers began to investigate methods for synthesizing diamond by decomposition of methane at high heat and low pressure. The method later became commercially viable in Japan (*Encyclopedia Britannica 2011 Standard Edition*: Synthetic diamond).

These experiments lend credibility to the possibility that diamonds exist in interstellar molecular clouds where there is low pressure but occasional high heat from star formation. It also shows that there could be a link between diamonds and supernova explosions.

Microdiamonds and graphite grains have been found in meteorites. There is some evidence that microdiamonds exist in the interstellar medium (Shaw 2006, page 140). Carbonado diamonds are black diamonds found in South America and Africa. They are used as industrial diamonds in drill bits. It is thought that carbonado diamonds were formed in interstellar clouds and then brought to Earth by some means. Microdiamonds that are older than the solar system are found in "presolar grains" in many meteorites. These extraterrestrial diamonds are thought to have formed in supernovas. The existence of microdiamonds in meteorites that are older than the solar system is a confirmation of the mantle accretion notion of the Planet Capture Hypothesis since it shows that bodies moving through space outside the solar system are picking up carbon material that has been transformed into diamond far beyond the confines of the solar system.

The link between organic material, hydrocarbons, methane, graphite and diamonds is clear: the remains of life can be transformed into the world's most precious gemstone. That these gemstones are found *inside* the Earth's mantle supports the Planet Capture Hypothesis since the carbon material from which the diamonds were formed appears to have been accreted from interstellar dust clouds or to represent bacterial colonies growing upon an ancient basalt surface later trapped by further accretion of the mantle.

Despite the foregoing discussion, if you are given a diamond ring, it would be unwise to say thank you for the lump of fossilized bacteria, since it could start you off on the wrong foot.

6. Signs of life on other planets

If unicellular life was accreted onto the Earth in its proto-planet stage as it passed through dark clouds, then it was also accreted onto other planets in the solar system, and moons that now orbit the gas giants.

Microbial life has not yet been directly detected on any other planet in the solar system, but there are intriguing indications of the presence or former presence of life on other planets. James Lovelock, British scientist (author of the Gaia Theory) proposed in the 1960s that life could be detected on another planet by analysing the composition of its atmosphere (easier than landing on the planet and taking samples). If the atmosphere contained two gases that would normally react together and deplete one another, there must be some living creature on the planet continually producing the gases.

Earth's atmosphere contains free oxygen (O_2) which is highly reactive alongside methane (CH_4) even though the two react together. Oxygen and methane coexist because oxygen is constantly generated by plants and methane is generated by anaerobic bacteria. Consequently, oxygen is a biomarker of life on Earth and methane could also be a biomarker. Lovelock saw the atmosphere as a complex system created by life.

The presence of methane is compared to that of oxygen in the atmosphere of the early Earth by Shaw (2006, page 220):

"Oxygen is a very reactive molecule and, without a continuous source to replenish it, would be used up and disappear from the atmosphere very quickly by reaction with surface materials. Hence detection of oxygen in the atmosphere either by the 0.76 μm absorption band of oxygen or the 9.7 μm absorption band of ozone would be a good biomarker for life. Similarly with methane, which again would be quickly destroyed were it not for a replenishment mechanism (on Earth, this is anaerobic organisms). Early organisms, such as the archaebacteria, would have respired CO_2 and converted this to methane so an early Earth would have a different atmospheric signature."

Shaw observes that both oxygen and methane must have a replenishment mechanism to exist in the atmosphere since they are highly reactive; on Earth methane is produced by anaerobic archaebacteria. This observation made by Shaw could be applied to any planet.

Astrobiologists searching for evidence of life on other planets and moons in the solar system have always taken the discovery of liquid water as an indication that the planet could support life. In Chapter 10 of Part I, I proposed that water is produced by life, especially anaerobic bacterial life. According to this view, water could be seen as a biomarker indicating the presence of life.

Scientists searching for signs of life on other planets have always favoured Mars as a place to look. Mars does not have a crust like the Earth; it has a surface of basalt rocks erupted from the mantle. The average temperature of the surface of Mars is very cold (218 K or -55 °C), although it reaches 20 °C in summer near the equator. Scientists are encouraged in their search by the evidence that Mars once had running water. There are outflow channels cut in the surface resembling arroyos in the desert (an arroyo is a channel in which water flows only occasionally). Mars has water in frozen polar ice caps and has subsurface permafrost.

In 2004 methane was detected in the atmosphere of Mars by telescopes on Earth and a satellite orbiting Mars. Methane gas would be broken down by the Martian atmosphere, so its presence indicates that it is being replenished by some means. A possible source would be volcanoes, but there are only dead volcanoes on Mars. Some scientists have proposed that the presence of methane in the Martian atmosphere is indicative that methane-producing methanogens are living on the planet Mars.

In Chapter 10 of Part I, I outlined how anaerobic bacteria such as methanogens convert hydrogen and carbon dioxide gases into methane gas and water. There is no requirement of light for this process and some internal heat inside a planet may be all that is necessary to provide the conditions for a bacterial community to thrive. The metabolic activities of anaerobic bacterial communities can lead to a massive build-up of methane and water. On Earth methanogens inhabiting swamps and marshes produce methane;

methanogens also inhabit the guts of animals. Volcanoes and subsurface natural gas reserves associated with oil fields are thought to be non-biological sources of methane; however, there is evidence now that subsurface bacterial communities may inhabit oil fields where they feed upon buried organic matter transforming hydrocarbons and producing methane and water.

Bacterial communities also produce other residues such as ammonia, hydrogen sulphide, amino acids and organic forms of phosphate.

Methane and water are surprisingly common on the surfaces and in the atmospheres of planets and moons orbiting the gaseous planets in the outer solar system. Ammonia is also common.

These are the surface compositions of planets in the outer solar system:

- The upper atmosphere of Jupiter is molecular, composed of hydrogen and helium, but also with traces of methane, ammonia and water.
- Saturn is similar to Jupiter.
- Uranus has an atmosphere mostly of hydrogen and helium, but with methane gas and ammonia clouds. The planet is covered with ice composed of water, methane and ammonia.
- Neptune has an atmosphere of hydrogen and helium with clouds of methane and hydrogen sulphide.

The moons or satellites of these giant planets show surprising diversity:

Among the 60 moons of Jupiter there is Europa, Ganymede and Callisto. Ganymede is covered by an 800 km thick mantle of water ice. Europa is also covered by water ice, but 10 km below the surface it is thought that there is a liquid salty ocean 200 km deep. The icy surface shows darker orange-brown areas and resembles an aerial photograph of the Arctic Ocean with ice floes crunching together. It is speculated that life inhabits Europa's salty ocean. Complex organic molecules with carbon-nitrogen bonds have been detected on Callisto.

Europa has almost no atmosphere, but the slight atmosphere it has consists almost entirely of oxygen. This oxygen could be produced by cosmic rays striking the icy surface and splitting water into hydrogen and oxygen, but if this is the case, the hydrogen produced is missing. It has been suggested that the lighter hydrogen has escaped into space, while the heavier oxygen remained. Ganymede also shows evidence of molecular oxygen in its atmosphere. The presence of molecular oxygen on Europa and Ganymede is intriguing since molecular oxygen has always been taken as an indication of the activity of photosynthesis performed by bacteria, algae or plants.

Of the numerous moons of Saturn, Titan stands out as having a thick atmosphere of 80 % molecular nitrogen with some methane. Complex organic molecules have been detected in its atmosphere including ethane, a gaseous hydrocarbon. Titan has deposits of dark hydrocarbons and methane rain. Recent evidence shows the presence of liquid lakes that are thought to be formed of liquid methane, although the surface temperature is –180 °C.

Of the 13 moons of Neptune, Triton is the largest. Triton has a thin atmosphere of nitrogen with traces of methane. Triton is one of the coldest places in the solar system (38 K at the surface); however, some nitrogen and methane gas appears to have evaporated in spots around the warmer south pole.

Pluto is covered with methane, nitrogen and carbon monoxide ices. These ices evaporate to form a thin atmosphere when Pluto approaches the Sun every 248 years.

Beyond our solar system the brown dwarf Gliese 229B which is a companion of the red dwarf Gliese 229A has an atmosphere containing methane.

The satellites of the Jovian planets have intriguing compositions. If it is life that produces methane, ammonia, hydrogen sulphide, water and complex carbon molecules, then anaerobic life has thrived, at least at one time around the Jovian planets.

In Chapter 13 of Part II, I proposed that the Jovian planets are failed stars. They resemble stars in that the bulk of their composition is hydrogen. If the Jovian planets are failed stars they would have started out as protostars formed within molecular clouds. At the outset they would have had a surface temperature of about 3000 K and given off red light. These gentle conditions would be favourable to life living in association with dust rings around the protostar. At this stage they may have captured some moons and the products of anaerobic life – water, methane and ammonia would accumulate on their surfaces.

Over time the small star failing to ignite nuclear fusion reactions would cool down and shrink in size becoming a brown dwarf. It may draw in methane and water with dust from the rings in orbit around it. At some point the miniature system could be captured by a larger star such as the Sun.

The continued cooling of the brown dwarf and its satellites may convert it into a frozen world in which methane remains due to its frozen state and water is only found as ice. The ices and gases found on the gas giant planets and their satellites in the outer solar system may be evidence of the former presence of life.

7. Basalts – source of elements for life

Basalt is black or grey volcanic rock originating from the Earth's mantle and rising to the surface as magma. Basalt is liquid at 1200 °C and solid below 1000 °C. When magma solidifies into lava it often contains air bubbles and has a frothy appearance.

Basalt formations on land include:
- Volcanic eruptions producing scoria, ash, cinder, breccias and lava flows. Cinder is often frothy and red in colour due to oxidized iron.

- Columnar basalt formed by the cooling of a thick lava flow (vertical lava flow within a volcano) with contraction producing joints between hexagonal columns. An example is the Giant's Causeway in Ireland.

- Continental flood basalts produced by episodes of volcanism. Extensive outflows form plateaus of basaltic rock often cut by rivers and waterfalls. Examples are the Deccan Traps in India and Parana Traps in Brazil.

- Greenstone Belts in the most ancient parts of continents are composed of Precambrian basalts erupted under water which have been uplifted above sea level by folding and faulting to form land. These basalts have been preserved by metamorphism causing the minerals to harden against weathering. Examples are the Barberton Greenstone Belt of South Africa and Swaziland and the Pilbara region of Western Australia.

Basalt formations of the ocean include:

- Ocean floor produced from the upwelling of magma at mid-ocean ridges. There are extensive ridge systems beneath the oceans.

- Pillow basalt – when basalt erupts under water it forms pillow shapes with a fine-grained core and glassy crust. Volcanic glass is associated with underwater eruptions.

- Oceanic islands are often formed from basalt, for example, Iceland and Hawaii.

Basalt is composed of the minerals olivine, pyroxene and plagioclase. Olivine has the formula $((Mg,Fe)_2SiO_4)$, clinopyroxene has the formula $((Ca,Mg,Fe,Al)_2(Si,Al)_2O_6)$ and plagioclase $(NaAlSi_3O_8)$.

By weight oceanic crust is composed largely of basalt with 50 % silica or silicon dioxide (SiO_2) and 4-15 % each of magnesium oxide (MgO), iron oxide (FeO), calcium oxide (CaO) and alumina (Al_2O_3). Metal sulphide minerals such as pyrite (FeS_2) are prominent at seafloor hydrothermal ridge axes (Edwards et al. 2003, page 180).

Types of basalt include tholeitic basalt which is poor in sodium, alkali basalt which is rich in sodium, boninite which is rich in magnesium and high alumina basalt when it contains over 17 % alumina (Al_2O_3). Basalt also often contains some iron-titanium oxides (TiO_2).

Plagioclase is composed of different types of feldspar. Feldspar is an aluminosilicate mineral. Alkali feldspar contains sodium as well as potassium. Potassium feldspar (K-feldspar) has the formula $(KAlSi_3O_8)$. There are also calcium/aluminium rich feldspars and rarely barium feldspars with the formula $(BaAlSi_3O_8)$.

Basalts weather quickly releasing calcium, sodium and magnesium and other elements to the environment.

Microbial activity

Oceanic crust is made of basalt on average 7 km thick covered by varying depths of sediments. Where basalt emerges at mid-ocean ridges the sea floor is much higher than in the vicinity of oceanic trenches; the ocean floor becomes progressively deeper away from the central crest of mid-ocean ridges. Ocean floors are young in comparison with the sedimentary strata of continents.

By contrast continental crust is on average 35 km thick and up to 3.8 thousand million years old. The basalts on land appear to only be conserved if they are metamorphosed by heat converting them into harder rock known as serpentine.

A possible explanation for lack of very old basalts and the thinness of oceanic crust compared to the roots of continents could be that basalts are constantly being consumed by microbes. Lithotrophic modes of metabolism would convert basalt rock into water, gases, organic matter and altered minerals. The effect would be to free elements to the environment which are the basis to life and the saltiness of the sea.

The porosity of basalt – it often has lots of bubbly holes in it – would give access to microbes living upon it. If microbial action continued over geological time scales, basalt rocks would not accumulate to great thicknesses where bacteria had access to them and would not be old in comparison with sedimentary strata.

At the bottom of the ocean there is no light, but bacteria living by chemosynthesis do not require light since they use chemical energy. Magma chambers beneath mid-ocean ridges are a source of heat to the surrounding rocks – at a certain distance from mid-ocean ridges bacterial metabolisms would have the necessary heat to work at maximum capacity. Corrosion features on underwater volcanic ridge system basalts suggest the presence of microbial communities.

Iron-oxidizing lithotrophs

Katrina Edwards et al. (2003) conducted a study of endolithic (rock-hosted) microbial communities on the Juan de Fuca Ridge axis off the coast of north western America. This study followed on from a study by Furnes and Staudigel (1999) [15] in which textural observations and carbon isotope measurements suggested that up to 75 % of the uppermost ocean crust is microbially altered. Other studies opposed this view considering that alteration of the rocks was probably mediated by physical or chemical processes. The emerging picture is, however, that ocean crust is altered by microbial activity, but the amount of this activity has not been ascertained due to sampling problems at the bottom of the ocean.

The Edwards et al. (2003) paper describes how samples were taken of microorganisms associated with the weathering of basaltic glass and sulphide minerals in the vicinity of hydrothermal vents and ridge flanks. This upper ocean crust was of less than 10 million years old.

The data shows that an unexpected diversity of microorganisms directly participate in rock weathering at the seafloor contributing to rock, mineral and carbon transformations.

The uppermost 200-500 metres of basaltic ocean crust is characterized by high permeability which facilitates the circulation of large quantities of seawater. Basaltic rocks react with oxygenated deep-sea water to form secondary minerals including iron-oxyhydroxides, micas and clays which fill fractures and void space in the crust. These reactions also maintain the chemical composition of seawater.

The low-temperature oxidation reactions release energy but do not occur easily. This provides an opportunity for microorganisms to use this chemical energy for metabolic growth. It was found that the bacteria inhabit pits in the basaltic rock and cover the surface with iron oxides as a crust. The crust prevents the non-biological alteration of the minerals beneath by direct contact with oxygenated seawater. This allows the bacteria to harness the energy released from the oxidation of ferrous iron for growth.

A wide diversity of autotrophic iron-oxidizing bacterial strains were found by Edwards et al.. They were obligate lithotrophs capable of growth with Fe^{2+}, and no other electron donor. Most of the strains

[15] Furnes, H. and H. Staudigel (1999) Biological Mediation in Ocean Crust Alteration: How Deep is the Deep Biosphere? *Earth Planet.Sci.Lett* Vol.166, pages97-103.

were Proteobacteria with no known autotrophic relatives, although they had heterotrophic relatives such as *Marinobacter sp.* There were also sulphur-oxidizing bacteria.

Manganese-oxidizing bacteria

The role of metal-oxidizing bacteria in the weathering of submarine volcanic rocks was further investigated by Alexis Templeton et al. (2005).

The study involved the collection of oxidized pillow basalts from the cold, outer flanks of Loihi Seamount, an active submarine volcano beside the island of Hawaii and at the older lava flows of South Point and Kealakekua Bay. The highly vesiculated basalts were collected at 1100-1715 metres depth where there is no light and the temperature is between 1.7 and 2.5 °C. The basalts were aged between 8 years and over 300 years. Manganese-oxidizing bacteria were cultured from these recent weathered basalts.

Manganese is distributed throughout the deep ocean as $Mn(II)$ in basaltic rocks and hydrothermal fluids, and as $Mn(III)$ and $Mn(IV)$ oxides in seawater particulates and deep-ocean ferromanganese nodules and crusts.

Low temperature basalt weathering under water-dominated conditions typically results in the large-scale mobilization of calcium, magnesium, silicon, aluminium and sodium from the volcanic rocks and the formation of oxidized secondary phases such as $Fe(III)$-oxyhydroxides and clays within alteration layers.

The potential for chemolithoautotrophy involving $Mn(II)$ oxidation in hydrothermally active regions such as seamounts and ridge systems and other areas of basaltic crust is untested. There are no reliable criteria for distinguishing biogenic manganese oxides from oxides formed abiotically. However, manganese oxides intermixed with iron oxides are found as dark coatings on the glassy margin of pillow basalts in association with orange microbial mats.

SEM imaging showed that the natural basalt surfaces were covered with microbial cells of diverse morphology. Cultures taken from oxidized pillow basalts formed numerous, diverse heterotrophic $Mn(II)$-oxidizing bacterial colonies. The isolates included Gram positive bacteria, Actinobacteria and Proteobacteria. In association with sulphur minerals the bacterium *Sulfitobacter* was isolated. *Sulfitobacter* can grow lithotrophically on sulphite; it appears to fix CO_2 by growth on reduced sulphur compounds.

Abiotic Mn(II) oxidation is very slow since it requires a high activation energy. Manganese oxides on basalt surfaces especially young glasses are likely to be indicative of microbial metal-cycling since enzymatic catalysis of Mn(II) oxidation increases the reaction by up to five orders of magnitude faster than abiotic (non-biological) processes (Templeton et al 2005, page 128).

Proteobacteria are a metabolically diverse group of chemoorganotrophic bacteria. They form red pigmented colonies.

It is interesting that it was found that the Proteobacteria strains *Methylarcula* and *Sulfitobacter* are rod shaped when suspended in liquid and take on a coccoidal form when attached to a surface. They produce exopolysaccharides on a surface. Mn(II) oxidation is only performed on plates, not in liquid suspension and only performed by coccoidal, not rod-shaped bacteria.

This observed conversion of morphology is a reason for not classifying bacteria by shape. Traditionally bacteria have been classified in terms of their three basic shapes: spirillum (spiral or helical-shaped), rod-shaped (these include short rods and long filaments) and cocci (round or spherical-shaped found as clusters or chains) (Mader 2007, page 371). In Chapter 6 of Part I I classified bacteria according to metabolic type.

Templeton, Staudigel and Tebo (2005) suggest in the paper that there could be a possible alternation between lithotrophic and heterotrophic modes of metabolism for *Marinobacter* and *Halomonas sp.*. Colonies often grow quicker when a carbon source is present suggesting a mixotrophic mode of growth with use of the carbon source for biosynthesis and Mn(II) oxidation for energy production. *Marinobacter* and *Halomonas* isolates grow under an extremely wide range of conditions.

There is the possibility that Mn(II)-oxidizing bacteria are directly dependent on Fe(II)-oxidizing bacteria. Fe(II)-oxidizing bacteria are likely to be initial colonizers of basalts with their autotrophic growth and primary carbon fixation leading to an accumulation of organic material on basalt surfaces. This would stimulate the activity of heterotrophic Mn(II)-oxidizing bacteria which use organic material as a carbon source for biosynthesis and metal oxidation to obtain energy by respiration. The activity of Mn(II) oxidation would reduce the oxygen level within the microbial biofilm. Lithotrophic Fe(II)-oxidizing bacteria often do not tolerate high oxygen levels, and so their activities would be enhanced by the activities of Mn(II)-oxidizing bacteria which reduce the oxygen levels.

Templeton, Staudigel and Tebo conclude that such synergistic interactions between lithotrophic and heterotrophic bacteria during the colonization of young basalts may be essential in the generation of ferromanganese crusts. Heterotrophic Mn(II)-oxidizers may be critical members of the biofilm microbial community in terms of their role in affecting the rates of volcanic glass dissolution and biomineralization (Templeton et al. 2005, page 137). The evidence suggests that Mn(II)-oxidizing bacteria are ubiquitously associated with weathered basalts.

Conclusion on basalts
The weathering of basalt rocks releases elements to seawater and leads to the formation of secondary minerals including iron-oxyhydroxides, micas and clays. The elements contained in basalt minerals include magnesium, iron, calcium, sodium, aluminium, potassium, manganese, barium and titanium. The mineral bulk is composed of silicate (SiO_2). With the exception of aluminium, barium and titanium the elements contained in basalt are required by life for the building of structures and physiological systems.

It is my contention that were it not for microbial action upon basalts, these elements would not be available on Earth to other forms of life. Bacterial lithotrophs appear to be the key to the initial decomposition of minerals transforming them into other types of mineral. Altered minerals such as clay are consumed by multicellular animals such as worms passing the required elements into the ecological chain of life. Clays are also the substrate to plants providing minerals, while mineral particles in seawater are the fertilizers for phytoplankton.

Many of the elements contained in the cells and structures of living organisms, the elements of sedimentary rock strata and the sodium in the salt of the sea had to come in the first instance from the minerals of Earth's mantle. The minerals of the mantle are exposed on Earth's surface by volcanic eruptions on land and under the oceans. The consumption of erupted basalts by bacteria on geological time scales could have enormous consequences for the environment and even be planet-changing.

8. Sources of water on Earth

In Chapter 10 of Part I, I presented the Life Produced Water Hypothesis and looked in detail at the microbiology of water. Water is the by-product of many types of bacterial metabolism. I proposed that life has caused a net accumulation of water on Earth.

The accumulated water on Earth in oceans, lakes and rivers as well as in glaciers, and falling as rain upon land surfaces, has made the Earth a habitable place.

In this chapter on Earth as a planet, I am going to look again at all the possible sources of water on Earth.

Anaerobic phase of planet Earth

The first and most obvious source of water on Earth would, according to the idea that Earth's mantle was formed by the accretion of interstellar dust grains, come from the ice mantle of the dust grains. We have seen that interstellar dust clouds are full of water ice. It has been proposed that this water was produced by anaerobic metabolism practiced by extraterrestrial forms of bacterial life. Thus, the accreted silicate minerals of Earth's mantle would be associated with abundant water.

Secondly, the basalt rock surface of the early Earth when it only had a mantle and no crust would have been inoculated with anaerobic bacteria literally eating the rock and converting it into water.

This anaerobic phase of early Earth lasted for over 2000 million years between 3.8 and 1.8 thousand million years ago. If the proto-planet phase of the Earth, before capture is counted, the anaerobic phase may have gone on for as long as 4000 million years in total. The addition of water drop by drop by anaerobic microbial metabolism for this length of time resulted in the formation of oceans on planet Earth.

Combustion
Apart from metabolism, the other major contributor of water is through the combustion of organic matter.

Subduction volcanoes Most of the active volcanoes on Earth (80 %) are subduction volcanoes arising from the melting of slabs of oceanic crust that have been subducted into the Earth's mantle at oceanic trenches (this will be examined in Chapter 21, section 3 of this book).

Subduction volcanoes release large quantities of carbon dioxide and water vapour into the atmosphere. It is thought that the gases of these volcanoes and the water vapour are produced by the combustion of organic matter contained in marine sediments which form part of the subducting crust.

An indication that the gases blown out by these volcanoes come from the combustion of organic matter is that the magma from subduction volcanoes gives rise to andesite lava which is silica-enriched lava. The silica component is likely to come from the melting of marine sediments containing the silica shells of marine organisms.

Hydrothermal vents Another prevalent type of volcanism on Earth is that associated with mid-oceanic ridges. The underwater lava flows of these ridges are associated with hydrothermal vents within a zone 1-2 km wide. Hydrothermal vents resemble chimneys from which plumes of hot water emanate at 350 °C loaded with sulphides of zinc, iron, cobalt and copper and reduced gases such as hydrogen sulphide. The vents are sometimes called black smokers because of the concentration of metal sulphides which blacken the water.

The current explanation for hydrothermal vents is that cold sea water penetrates the crust through cracks and re-emerges as hot water heated by the proximity of magma chambers. Lamb & Sington write (1998, page 60),

"It has been estimated that in a few tens of millions of years, the entire volume of the world's oceans is flushed through cracks in the ocean floor."

I propose that rather than sea water disappearing down cracks and the *same* water re-emerging after having been heated, the water of hydrothermal vents is being produced by the combustion of organic matter trapped in marine sediments below fresh flows of basalt lava. Combustion of organic matter due to contact with or proximity to molten magma would produce carbon dioxide and water, as well as the hydrogen sulphide given out by hydrothermal vents. Whereas subduction volcanoes blow out steam at the surface, the atmospheric pressure at the bottom of the ocean means that water emerges as liquid from the vents, rather than as water vapour.

In Chapter 10 I proposed that rocks in the vicinity of hydrothermal vents that are warm, but not hot (with a temperature below 100 °C) may harbour subsurface bacterial communities whose metabolism contributes to the water emerging from hydrothermal vents. This would be another source of subsurface water produced beneath the ocean floor.

Hydrothermal vents have their own communities of animal life living on the seabed consisting of clams and worms which have evolved a symbiotic relationship with sulphur-oxidizing bacteria. The bacteria harboured in the animal's body are able to exploit hydrogen sulphide as an energy source in the total darkness of the ocean floor. The sulphides are transformed into sulphates, and sulphuric acid.

Combustion of marine sediments would not only explain the production of water at hydrothermal vents, but it would also explain the absence of ancient sediments in oceanic crust. Of course, only the organic part of the sediments combusts to form water and gases; the silica from shells of marine organisms is redeposited by hydrothermally heated waters as secondary silica and the calcium carbonate from other shells is redeposited as secondary calcite. Secondary silica deposited by hydrothermal water forms veins of chalcedony or large crystals of quartz called rock crystal within other rocks; calcite is also found as veins within other rocks.

Aerobic phase of planet Earth
Under current conditions there is evidence that aerobic
chemolithotrophs living on ocean floors continue to produce water by
transformations of basalt rock. This has been going on for at least 1.8
thousand million years. Secondly, there appears to be a vast
subsurface biosphere where anaerobic bacteria still eek out a living on
buried organic matter contributing to water on Earth.

Ferromanganese nodules and crusts Basalts erupted in the
vicinity of mid-oceanic ridge systems are weathered to produce
ferromanganese nodules and crusts that are so abundant on the ocean
floor that they could be mined if it were economically viable to do so.

It is now becoming apparent that the weathering of basalts may be
largely mediated by chemolithotrophic bacteria (as explained in
section 7). Basalt lava appears to first be colonized by iron-oxidizing
bacteria that fix carbon dioxide in seawater using energy obtained
from oxidizing iron contained in basalt with oxygen in seawater.
These bacteria are obligate lithotrophs – they 'eat' rock and build up
cell material from inorganic matter. The basalt lava is then colonized
by manganese-oxidizing bacteria which are generally heterotrophs
living on organic matter as a carbon source, but using the oxidation of
manganese to obtain energy.

Nodules at the bottom of oceans contain significant quantities of
manganese, iron, copper, nickel, titanium and cobalt, as well as traces
of other metals. Aerobic metal-respiring bacteria may be the
producers of these ores by removing the reduced or lower oxidation
state metals from basalt rocks, oxidizing them and concentrating them
into secondary deposits. Ores are mostly deposits of oxidized metals.

Iron bacteria which are aerobic chemolithotrophs convert ferrous to
ferric iron or hydroxides of iron to a higher oxidation state in the
presence of oxygen. Other metal-respiring bacteria also oxidize metals
in the presence of oxygen to obtain energy. The point to be made here
is that metabolism involving the oxidation of metals is accompanied
by the production of water.

The equation for the production of water by iron bacteria given in Chapter 10, section 4 of Part 1 is as follows:

$$2Fe^{2+} + 2H^+ + \tfrac{1}{2} O_2 \rightarrow 2Fe^{3+} + H_2O$$

Oil reserves and methane seeps Oil comes from the anaerobic decomposition of Precambrian accumulations of cyanobacteria and unicellular algae. Oil wells are inhabited by anaerobic bacteria breaking down organic matter to form kerogen and various hydrocarbon compounds. Archaebacteria known as methanogens generate methane – the natural gas associated with oil wells and water in large quantities by metabolic conversions.

Methane seeps are found along continental margins where buried sediments containing organic matter may harbour communities of anaerobic methanogens. Methane seeping into the oceans from this source would also be accompanied by a spring of water.

Methanogens can also convert hydrogen released from basalt rocks and atmospheric carbon dioxide into methane and water. Carbonates can replace carbon dioxide as a carbon source. In this way rock can literally be converted into water.

Freshwater springs On land springs of fresh mineral water arise from aquifers of freshwater where the water table reaches the surface. Aquifers are permeable sedimentary rocks that form reservoirs for water – sandstone stores water around the grains of sand, while limestone is riddled by underground channels that contain water.

Much of the water contained in aquifers filters down into the ground when it rains. The water re-emerges from underground as springs which form streams which form rivers going to the sea. Clouds form above the sea and are blown over land bringing rain. This is the hydrological cycle.

I venture to propose, however, that a certain proportion of fresh underground water is produced by aerobic chemolithotrophic bacteria.

Hydrogen is produced by the anaerobic fermentation of organic matter in sediments at the bottom of lakes or by buried organic matter. Above the anaerobic zone there may be an aerated zone where oxygen is available. Hydrogen bacteria are able to oxidize hydrogen by reacting it with oxygen using the hydrogenase enzyme. In this way hydrogen bacteria obtain energy in the absence of light underground and either fix carbon dioxide from the atmosphere to make cell material or live on organic matter as a carbon source.

As explained in Chapter 10, section 4 of Part I, hydrogen bacteria such as *Aquifex* – the 'water-maker'and *Hydrogenobacter* produce pure water from the gases hydrogen and oxygen. Many of these bacteria are thermophiles thriving at high temperatures; *A. pyrophilus* has an optimum temperature of 85 °C. Heat tolerance makes hydrogen bacteria well adapted to the subsurface environment.

Another set of aerobic chemolithotrophs are the methanotrophs which are bacteria that oxidize methane under aerobic conditions to obtain energy. This occurs where methane is not spontaneously converted to carbon dioxide and water.

Thus, in conclusion to this section on sources of water on Earth, I propose that most of Earth's water came from the accretion of interstellar dust which is associated with water ice and from the conversion of basalt rocks by anaerobic bacteria during the very long Precambrian phases of planet Earth.

I believe that the amount of water on Earth is continually rising through:

- The conversion of subducted or buried organic matter into water and gases (carbon dioxide) by combustion. Subduction volcanoes blow the water off as steam while at mid-ocean ridges liquid water emerges from hydrothermal vents.

- The activities of chemolithotrophic metal-oxidizing bacteria under aerobic conditions. The conversion of metal compounds contained in basalt rock into oxidized metal ores lying on the seabed as nodules and crusts is performed by iron bacteria and manganese-oxidizing bacteria with each metabolic conversion being accompanied by the production of water. The extent of weathering of basalt rock formations under the sea and the quantity of metal-rich nodules on the seabed is an indication as to the importance of this mode of water production from basalt.

- The conversion of ancient buried organic matter into water and gases (methane) by anaerobic bacteria, principally methanogen Archaebacteria. Most of this organic matter was accumulated during the Precambrian era. Its conversion into kerogen and oil is accompanied by the production of methane gas and abundant water. When methane is further oxidized in an aerobic atmosphere by aerobic methanotrophic bacteria or by combustion in the burning of fossil fuels, it converts into carbon dioxide and more water. Thus, elements trapped as buried organic matter from the early Earth era become atmospheric gases and water when released in the present era.

- Freshwater on Earth only represents 2 % of the hydrosphere. I propose that a certain percentage of groundwater is produced by aerobic chemolithotrophic bacteria from the gases hydrogen and oxygen. The principal producers of this fresh spring water would be hydrogen bacteria that can live in darkness in the subsurface environment and tolerate reasonably high temperatures.

Therefore, a rise in the amount of water in the oceans could ultimately be attributed to the combustion of organic matter in marine sediments; the metabolism of anaerobic methanogen bacteria living on buried organic matter; and the metabolism of aerobic metal-respiring bacteria living on basalt.

Conclusion

The Planet Capture Hypothesis has taken us through an extra-solar phase in the formation of the Earth involving the accretion of the mantle, and the inoculation of unicellular life as the proto-planet passed through interstellar dust clouds. There may have been life before light on the surface of the Earth.

The icy worlds of the outer solar system appear as the frozen remains of communities of microbial life which once thrived around proto-stars.

It is envisaged that the Earth started out as a planet with a dry basalt surface resembling other terrestrial planets which have the minerals of the mantle as the surface. It is proposed that the initial transformation of this surface was by microbial action continued on geological time scales.

The metabolism of chemolithotrophic bacteria living on basalt rock produced water; the combustion of buried organic matter also produces water. At the same time, these chemolithotrophic bacteria convert the minerals in basalt into clays and micas.

Although the interior of the Earth is just below our feet, it is not well-known and is still the subject of much controversy. Likewise the ocean floor is not far away, but it has not been fully explored. There is still much to be discovered about planet Earth.

Bibliography

Edwards, Katrina J., Wolfgang Bach & Daniel R. Rogers (2003) Geomicrobiology of the Ocean Crust: A Role for Chemoautotrophic Fe-Bacteria. *Biology Bulletin* Vol. 204, pages 180-185. (April 2003).

Encyclopedia Britannica 2011 Standard Edition: Chemical element: the Earth's core; Diamond; Earth: the interior; Feldspar; Hydrosphere; Industrial diamond; Iron processing; Kelvin, William Thomson, Baron; Nickel processing; Ocean; Olivine; Pyroxene; Steel; Synthetic diamond.

Geikie, Archibald (2009) *Charles Darwin as Geologist: The Rede Lecture, given at the Darwin Centennial Commemoration on 24 June 1909* Cambridge University Press

Hyndman, Donald (1985) *Petrology of Igneous and Metamorphic Rocks* 2nd Edition McGraw-Hill

Ince, Martin (2007) *The Rough Guide to the Earth* Rough Guides Penguin Group

Lamb, Simon & David Sington (1998) *Earth Story: The Shaping of Our World* BBC Books

Prescott, Lansing M., J.P. Harley, and D.A. Klein (1999) *Microbiology* 4th edition McGraw-Hill

Shaw, Andrew (2006) *Astrochemistry: From Astronomy to Astrobiology* John Wiley & Sons Ltd.

Taylor, Stuart Ross (1998) *Destiny or Chance: Our Solar System and its Place in the Cosmos* Cambridge University Press

Templeton, Alexis S., Hubert Staudigel & Bradley M. Tebo (2005) Diverse Mn (II)-Oxidizing Bacteria Isolated from Submarine Basalts at Loihi Seamount. *Geomicrobiology Journal* Vol. 22, pages 127-139.

Zeilik, Michael (2002) *Astronomy: The Evolving Universe* 9th edition Cambridge University Press

CHAPTER 19

CATASTROPHES STRIKE THE EARTH

Introduction

This chapter focuses on the effects on Earth of capture events in the solar system according to the Planet Capture Hypothesis presented in Chapter 17. In this chapter the disturbances provoked by planet capture events in the solar system are presented as the causes of mass extinctions on Earth.

By linking the evidence of mass extinctions in the fossil record of Earth's sedimentary rock strata and the signs of recent capture of certain terrestrial planets I came up with a capture scenario presented in Section 7 of Chapter 17. This chapter expands on the idea of planet capture from the point of view of Earth's surface features and Earth history.

A major disturbance could affect the axis of Earth's rotation and its orbit; it could produce gravitational waves affecting the interior of the Earth; gravity could cause the oceans of the Earth to produce giant tides; and it could cause small bodies in the solar system to take on a collision path with Earth.

The catastrophes on Earth have included: changes to seasonality leading to ice ages; episodes of volcanism lasting millions of years; tidal waves sweeping across land masses; and asteroid strikes. Through all of this the surface of the Earth has been transformed and life upon the face of the Earth has undergone major transitions which mark the eras and periods of the geological record.

1. Mass extinctions

Life on Earth is adapted to regular climatic cycles. The biological mechanisms of plants and animals rely to a large extent on a high degree of regularity in the yearly cycle. This is especially marked in animals living in temperate zones whose migration patterns and reproduction is determined by the seasons. It is equally true of plants in temperate zones whose growth patterns and seed production is determined by the seasons. The plants and animals living on the equator rely on a constant high temperature that never drops below a certain level, while the animals living in arctic conditions rely on constant cold conditions and the maintenance of a cold climate ecosystem.

Any changes that affect climate can have huge repercussions on the ecosystems of the Earth. We are now seeing how a rise of a few degrees in the Earth's global temperature may be caused by carbon dioxide emissions from the burning of fossil fuels. Global warming could have a drastic effect on the distribution of species and lead to the extinction of some species, as well as causing extreme weather conditions that have a huge effect on human life. If increases in greenhouse gases raising the Earth's atmospheric temperature by only a few degrees can have such wide-ranging consequences, what consequences would an extraterrestrial disturbance have?

It came to me in 1994 that while Earth's regular planetary cycles maintain life on Earth, any disturbance in them also has the potential to destroy life. If planets have been captured by the Sun after the Earth's capture, these events in the solar system would possibly have massive effects on life on Earth.

Planets are rotating. It seemed to me that gravitational forces produced in a capture event could alter the angle or direction of motion without changing the motion itself. Thus, the axis of rotation of a planet could be changed, while the rotation itself continued.

A simple count of the number of mass extinctions on Earth seemed to tally with the number of potentially captured planets after Earth's capture into the solar system. Of course, the greater the knowledge of the subject matter, the more complicated the picture becomes. Even so, I have not lost sight of my original simple picture connecting extinctions on Earth with planet capture.

The major extinctions on Earth
Geological timescale is divided into three main eras. The boundaries between the eras are marked by mass extinctions. The Palaeozoic era of ancient life is divided from the Mesozoic era by the late Permian mass extinction 245 to 225 million years ago. The Mesozoic is divided from the present Cenozoic era by the late Cretaceous mass extinction 65 or 66 million years ago. The eras are divided into periods each marked by a different collection of flora and fauna forming an ecosystem under different environmental and climatic conditions.

Traditionally the period prior to the Cambrian has been called the Precambrian, although some now divide this time span into periods with other names. The Precambrian spans a period of 2930 million years. All other periods span 32 to 70 million years.

Precambrian fossils start with stromatolites which are fossil cyanobacterial mats dating to 3500 million years old. Passing by the appearance of acritarchs which are fossils of single-cell protist protozoans at least 1800 million years ago, the Precambrian ends with seas colonized by cnidarian sea pens, corals and jelly fish, protist multicellular seaweeds and sponges between 620 and 570 million years ago. At 570 million years ago the Cambrian period of the Palaeozoic era starts with its famous arthropod trilobites, as well as shell-bearing animals and worms.

There appears to have been at least one major event during the Precambrian 600-650 million years ago which would have affected the unicellular life living on planet Earth at the time.

The geological record that spans the time in which multicellular organisms have lived on Earth during the last 600 million years has evidence of five mass extinctions (Wilson 1988, page 29). Mass extinctions are when many species simultaneously disappear from the fossil record.

The mass extinctions are as follows:

- Late Ordovician 440 million years ago: the brachiopods and trilobites were very much reduced in numbers.

- Late Devonian 365 million years ago: again brachiopods and trilobites were affected. Many ammonites, gastropods, some corals and fish groups died out.

- End of Permian 245 million years ago with a further episode of extinction 225 million years ago: this was the end of the trilobites and placoderm fishes. In the oceans many foraminiferans and corals died out, as well as warm-adapted marine animals. On land pelycosaurs, amphibians, reptiles and therapsids died out and some plants.

- Triassic 210 million years ago: extinction of dominant land vertebrates (dicynodonts and rhynchosaurs) as well as echinoderms, bryozoans, brachiopods, mollusks and fishes in the sea.

- Late Cretaceous 66 million years ago: extinction of the dinosaurs, pterosaurs and most marine reptiles. Foraminiferans and ammonites went extinct in the oceans, and magnolias became extinct on land (although magnolias are absent from the fossil record for the entire Cenozoic era, they are now common in gardens).

In mass extinctions 20-25 % of families and up to 50 % of species died out throughout the world. However, in the end of Permian extinction 50 % of families and 95 % of species died out. This brought an end to the era of ancient life, the Palaeozoic. It took 10 million years for faunas to return to previous levels.

There were also lesser extinctions that occurred in the late Cambrian, at the Jurassic-Cretaceous boundary, in the late Cretaceous and in the late Eocene.

If one includes the event of the Precambrian that would have affected unicellular life, the colossal end of Permian extinction, four other mass extinctions of multicellular life forms and four minor extinctions, there may be ten events to be accounted for.

I am going to link the Planet Capture Hypothesis presented in Chapter 17 to mass extinctions on Earth with the following hypothesis:

The **Mass Extinctions on Earth Hypothesis** can be stated thus:

The mass extinctions on Earth are evidence of cataclysmic events provoked by disturbances in the solar system involving planet capture. Gravitational interaction between the Earth and a captured planet or large body passing near to the Earth have caused changes in the axis of rotation of the Earth leading to ice ages; earth tides leading to episodes of volcanism; and the gravitational pull on the oceans has produced giant tidal waves. Giant tidal waves were also produced by asteroid strikes to the oceans on Earth during certain periods. Deposition of sediments by tidal waves was the main agency of mass extinctions and the concurrent formation of deep fossil beds.

2. Disturbances in the solar system

In the last section I suggested a link between the mass extinctions of life on Earth and disturbances in the solar system. The notion that catastrophes on Earth can be provoked by disturbances in the solar system has already found some favour in science since one mass extinction is widely accepted as being attributable to a disaster originating from direct impact with an asteroid dislodged from the asteroid belt.

The Asteroid Theory
Sixty six or 65 million years ago at the end of the Cretaceous period, an event brought an end to the Mesozoic era of dinosaurs living on land. It also wiped out many animals living in the oceans. Among the large animals living in the oceans were marine reptiles such as Pliosaurs and Plesiosaurs. Among the small animals were mollusks such as ammonites, belemnites and some bivalves. The Echinoderm crinoids and planktonic protists such as the foraminiferans, radiolarians, coccolithophores and diatoms were also decimated at this time.

Flowering plants survived and small mammals and birds, and these took over the world in the following era.

The now classic cause attributed to bringing about the end of the dinosaurs is an asteroid strike on Earth. The Asteroid Theory was proposed by Walter Alvarez, an American geologist and his father Luis Alvarez in the early 1980s. The theory is that the impact of a 10-20 kilometre diameter asteroid striking the Earth would throw up dust that would obscure the Sun and set off a disastrous chain of events including global cooling that would cause many species to go extinct.

Evidence for an asteroid strike is found in the K-T boundary. The rare metal iridium is abundant in this clay layer worldwide, but is generally rare in other strata of the Earth's crust. Iridium is abundant in asteroids, meteorites and the Earth's core. The source of iridium in the K-T boundary is thus thought to come from an asteroid. A large impact is also suggested by shocked quartz in this white clay layer. The impact crater is now thought to exist beneath the Yucatan Peninsula in the Caribbean.

I think that the evidence for an asteroid strike at the end of the Cretaceous period is convincing. I wish to take the logic of disturbances in the solar system much further.

Capture scenario
In Section 7 of Chapter 17 I proposed that the Jovian planets were captured by the Sun early on, while Mars and a planet orbiting where the asteroid is now located were captured before or at about the same time as the Earth.

The degree of inclination of the orbit of a planet to the plane of the solar system (or plane of the ecliptic) is used as an indication of the order of capture events. This is because a planet's orbit will conform to the plane of the solar system increasingly over time.

Jupiter, Saturn, Uranus and Neptune all have orbits which conform closely to the plane of the ecliptic. (The plane of the ecliptic should be defined by the orbit of Jupiter and not the orbit of the Earth because of its size and early capture). The orbit of Mars also conforms closely to the plane of the ecliptic.

I propose that Venus, Mercury and Pluto were captured after the Earth was captured and so produced events that affected the Earth. The orbit of Venus is inclined at 3.4 degrees to the plane of the ecliptic, the orbit of Mercury at just over 7 degrees and the orbit of Pluto at 17.2 degrees.

Disturbances in the solar system affecting the Earth could originate from various causes:

- Capture of a planet into the inner solar system crossing the orbit of the Earth to take up an orbit closer to the Sun (Venus, Mercury).

- Collision of a satellite captured by a giant gaseous planet with the planet itself or with another satellite in the outer solar system.

- Planet-sized bodies passing close by the Earth before being engulfed by the Sun (this could result in increased Sun-spot activity also).

- A massive collision with a planet which once orbited where the asteroid belt is now found beyond the orbit of Mars. The resulting pieces of asteroid were sent spinning in all directions.

This is the planet capture and mass extinction scenario in greater detail:

Venus – Precambrian event
I propose that the capture of Venus into the inner solar system was a relatively recent event. Venus' orbit is inclined at 3.4 degrees to the plane of the solar system and it shows signs of recent capture. I propose that the capture of Venus was the cause of a mass extinction which occurred 650 million years ago with an extensive Precambrian ice age.

Venus shows signs of being a very old planet in that it has slow rotation indicating that it has lost momentum over time (there is no perpetual motion) and it has no magnetic field indicating that its core has cooled down and solidified. Venus was captured into an almost circular orbit due to its slower velocity upon capture.

Despite being an old planet, Venus shows signs of recent capture in the extreme turbulence of its atmosphere and in its active volcanism.

The atmosphere of Venus is composed of 96 % carbon dioxide, 3 % nitrogen, some argon and traces of water vapour and oxygen at 90 times the Earth's atmospheric pressure (see footnote [16]). It has clouds thought to be composed of 90 % sulphuric acid mixed with water. This inferno of an atmosphere with fierce winds and cyclones surrounds a surface at oven temperature (470 °C). The brilliance of Venus which appears in the sky at sunrise and sunset as the 'morning star' and the 'evening star' comes from the swirl of its lethal clouds.

Venus has extensive high lands produced by an outpouring of lava from volcanoes. Taylor (1998, page 132) states that the volcanic episode took place a few hundred million years ago and the present surface of Venus is between 300 and 500 million years old. The large volcanoes on Venus have very steep slopes. This could be taken as an indication of recent eruption.

The cratering on Venus also gives the impression of being recent. Taylor (1998, page 131) writes, "The craters formed by meteorite impact on the surface of Venus are surprisingly fresh (........). Only rarely are they covered or entered by lava flows." The uneroded state of the craters on Venus indicates an age for them of between 300 and 700 million years (Taylor 1998, page 131).

The Precambrian event 650 million years ago caused extensive glaciation on a global scale, even bringing glaciers to places near the equator such as South Africa and Australia. This ice age brought an end to a world dominated by unicellular life forms and heralded the Palaeozoic era of multicellular plant and animal life. The large accumulations of cyanobacteria and algae became the raw material of today's oil fields.

---

[16] The Earth has its carbon dioxide mostly trapped as huge accumulations of limestone deposited by aerobic forms of life. This means that the carbon dioxide has not remained in the atmosphere as is now the case on Venus. Earth has 0.03 % carbon dioxide in its atmosphere, compared to 96 % in the atmosphere of Venus.

Mercury – end of Permian event

At some point Mercury sped into its position close to the Sun with a highly elliptical orbit that swings around the Sun. Mercury's orbit has an inclination of just over 7 degrees to the plane of the ecliptic.

Mercury is barely a planetary core with a thin mantle covering. Mercury retains a magnetic field which indicates a liquid core and that it is not a very old planet. Mercury's slow rotation is locked onto the Sun due to its proximity.

I am going to link the capture of Mercury to the massive end of Permian extinction that ended the Palaeozoic era. The end of Permian event appears to have been more than one event since it is dated to 245 and 225 million years ago, and changes also occurred 260 million years ago.

At the Permian extinction 95 % of species both on land and in the oceans were wiped out. The low-lying land surface with its swamp forests became buried later to be converted into beds of coal.

An episode of volcanic eruptions in Siberia followed the end of Permian event. Two hundred and fifty million years ago great volcanic lava flows known as the 'Siberian traps' erupted flood basalts. The flood basalts covered an area of 4.8 million square kilometres with lava 1.6 kilometres thick.

Asteroid collision – End of Cretaceous event

I propose that only 66 million years ago a large body, maybe even planet-sized, captured by the Sun collided with a planet orbiting beyond Mars. The massive collision created a ring of debris which is now the asteroid belt. Asteroids are essentially irregularly shaped pieces of planetary core. One asteroid sent spinning into the inner solar system collided with the Earth. Mars captured two asteroids – Phobos and Deimos – as moons. Other asteroids will have disappeared into both Jupiter and the Sun.

The remaining asteroids number about 5500. They have highly inclined orbits to the plane of the solar system. The largest asteroid Ceres has an orbital inclination of 10.6 degrees. The asteroids tend to have quite elliptical orbits and are all fast-spinning. The non-conformity to the plane of the solar system, elliptical and thus energetic orbits, and fast rotation denote a recent origin for the asteroid belt. Elliptical orbits that cross the path of the surrounding planets put asteroids on paths of capture or collision.

A massive collision in the asteroid belt zone would have produced a blast wave and direct asteroid impact producing extinction on Earth. There may have been several asteroid strikes. The end of Cretaceous event wiped out the dinosaurs on land, and mollusks such as ammonites and planktonic organisms in the seas. The event also provoked an episode of volcanism centred on India. The Deccan Traps in India date from 65 million years ago and erupted for over a million years.

In addition, the massive collision may have caused the rotational axis of Mars to start veering between zero and 60 degrees.

Engulfed by the Sun
The three eras of the geological record – the Palaeozoic, Mesozoic and Cenozoic eras have been divided from each other by mass extinctions. I have suggested capture events as possible causes of these major extinctions. The periods within an era have often also been divided from each other by extinctions. A number of smaller mass extinctions were described in the previous section.

Another possible cause of extinctions on Earth could be planet-sized bodies passing close by the Earth before being engulfed by the Sun. The size and gravitational attraction of the Sun makes being engulfed by the Sun extremely probable for any object entering the inner solar system. Whether a body collides with the Sun or goes into orbit around the Sun would depend upon its velocity and trajectory.

The staggered nature of the end of Permian extinction could be explained by the planet Mercury being accompanied by another small planet on almost the same trajectory, but which disappeared into the Sun leaving only Mercury in orbit close to the Sun.

Satellite collisions

The giant planets Jupiter, Saturn, Uranus and Neptune may have arrived with some satellites in orbit and captured others since their arrival in the solar system. Many satellites are the size of small terrestrial planets. In-coming bodies captured by the Sun but straying too near the Jovian planets could on some occasions have collided with the giant planets or with other satellites, and on other occasions taken up orbit around the giant planets as moons.

Among satellites that show signs of recent capture there are the following:

- Triton which has a retrograde orbit around Neptune that is highly inclined to Neptune's plane. Triton is spiralling into Neptune.

- Nereid has a highly inclined orbit around Neptune.

- Miranda has a highly inclined orbit around Uranus.

- Hyperion has a chaotic orbit around Saturn.

- Pasiphae, Sinope, Anake and Carme have retrograde orbits around Jupiter.

- Io orbiting Jupiter has very hot lava pouring from its interior – a sign of youth.

The collision of a satellite with a giant planet disappearing into it would send shock waves throughout the solar system. Shock waves on this scale, even though they come from the outer solar system, could provoke effects on Earth resulting in smaller mass extinctions.

As evidence of the real possibility of this happening, it is known that since Triton is spiralling into Neptune, it will one day collide with the giant planet producing shock waves with possibly devastating effects.

Further evidence of the possibility of a satellite colliding with a giant planet is the Great Red Spot in Jupiter's atmosphere. Extreme cyclonic turbulence in this area is suggestive of something having recently been engulfed by the gas giant.

Pluto

Pluto has the greatest non-conformity of all planets to the plane of the solar system; Pluto's orbit has an inclination of 17.2 degrees. According to the logic that I have employed in this section, this suggests that Pluto is the most recently captured planet into the solar system. With its companion Charon, Pluto careers past Neptune in a highly elliptical orbit.

Because Pluto is so far out in the outer solar system, I am not going to attribute Pluto's capture to any major hand in disasters on Earth.

In this section I have made a link between events in the solar system affecting the Earth as a planet and extinctions of life on Earth. The scenario offered is designed to give an illustration of the Mass Extinctions Hypothesis. Greater knowledge could, of course, lead to modifications in the scenario. It is open to investigation.

3. Ice ages

The rotation of the Earth sets the alternation of daylight and night. The length of a day – 24 hours is determined by the speed of rotation. The size of the orbit and the time taken to complete one orbit determines the length of a year – 365 ¼ days. The shape of the orbit in terms of eccentricity determines the light intensity from the Sun; this affects climate. The axis of rotation which tilts the poles towards or away from the Sun as it moves in its orbit produces the seasons.

The present tilt of the Earth is 23.27 degrees to the plane of the ecliptic (the ecliptic is the plane of Earth's orbit around the Sun). This produces the seasons which are felt more acutely in the temperate regions. The northern hemisphere experiences summer when the North Pole is tilted towards the Sun while the southern hemisphere experiences winter as the South Pole points away from the Sun. When the Earth reaches the other side of its orbit, the situation is reversed with the long nights affecting the northern hemisphere and the long hours of daylight enjoyed in the southern hemisphere.

If the Earth was upright in position with no tilt, the poles would still be cold due to the obliquity of the angle of the Sun's rays reaching them, and there would be climatic zones ranging from temperate at high latitude to tropical at the equator, but there would be no seasons. The duration of day would be 12 hours followed by 12 hours of night for every part of the globe. Under these conditions the northern and southern hemisphere temperate climatic zones would be uniformly mild in temperature, experiencing neither the heat of summer nor the cold of winter.

If the Earth was lying on its side like Uranus, it would experience almost 6 months of winter with permanent darkness followed by almost 6 months of summer with permanent daylight in each hemisphere alternately. Uranus has 42 years of night followed by 42 years of day on each side due to the size of its orbit around the Sun. The Sun-facing regions near the equator would experience mild temperatures throughout the year except for two hot seasons when the revolving equator was head-on to the Sun.

At the present tilt of the Earth, the Arctic Circle at just over 66 $^{\circ}$ latitude is defined as the zone where for one day each year the Sun does not set. This is near the 21st June and the Sun does not rise near the 21st December. The length of continuous day or continuous night increases northward from one day on the Arctic Circle to 6 months at the North Pole. There is also an Antarctic Circle around the South Pole which shows the same conditions, except they are opposite to the Arctic Circle.

Therefore, tilted planets have seasons and the degree of tilt will affect the amount of seasonality. An upright planet will have more evenly distributed temperature over the year with stable climatic zones. A planet lying in the plane of its orbit will have extreme variations in temperature in each hemisphere on a yearly basis, but the equator would show a more even temperature distribution between cool, mild and hot.

The tilt of the Earth means that the place where the Sun is directly overhead at midday moves from the Tropic of Cancer in the northern hemisphere to the Equator, and down to the Tropic of Capricorn, and back again over 12 months.

The Milankovitch Hypothesis

Milutin Milankovitch, a Yugoslav geophysicist proposed a theory in 1930 to explain the phases in the last ice age by changes to the Earth's tilt and orbit. Milankovitch was seeking an understanding of the glacial advances and glacial retreats that characterized the Pleistocene ice age which started 1.6 million years ago. The colder phases of the ice age have been interspersed with interglacial stages when temperatures in the northern hemisphere were higher than they are today.

Milankovitch first observed that it is the angle at which the Sun's rays hit the Earth that determines the amount of solar radiation reaching any one area. If the seasons changed such that the winter became long and the summer short, there would be a build-up of ice due to freezing temperatures in winter not compensated by melting of the ice in summer. A long series of cool summers would result in the formation of glaciers and the advance of glaciers. Milankovitch studied the occurrence of glacial maxima and drew up curves showing the variation of incident solar radiation over the last few hundred thousand years in the northern hemisphere.

Milankovitch proposed that there is a 41 000 year cycle of glacial advances whose ultimate cause is changes in the tilt of the spin axis of the Earth. This is combined with a wobble in the Earth's axis that occurs on a 22 000 year cycle. These changes in the tilt affect the incidence of solar radiation to different parts of the Earth and the intensity of the seasons. Secondly, Milankovitch linked ice ages to changes in the eccentricity of the Earth's orbit occurring on a 92 000 year cycle with the severest period of an ice age coinciding to a more circular orbit.

The wobble of the Earth's axis is called the precession of the equinoxes. The wobble resembles the motion of a spinning top. A top will wobble about its axis as it slows down (the Earth's rotation is slowing down). This motion makes the Earth's poles describe a circle in the sky over a period of 19 000 to 23 000 years. The details of how the precession of the equinoxes affects the amount of heat arriving at different latitudes in different seasons had, in fact, been worked out by James Croll in the 1860s. Croll was a self-taught Scotsman – his name, however, does not often get a mention.

114

In the 1970s scientists correlated oscillations in the oxygen isotope record in deep sea cores with the solar radiation cycles calculated by Milankovitch. They concluded that 60 % of climatic fluctuation can be explained by the curves of solar radiation calculated by Milankovitch.

The obliquity of the Earth's spin axis changes from 21.8 degrees to 24.4 degrees over a period of around 41 000 years. When the tilt is greater, the seasons get more severe, with hotter summers and colder winters.

Changes in the eccentricity of the Earth's orbit between more elliptical and more circular are also observable. The eccentricity of the Earth's orbit varies over two cycles, one of 100 000 and one of 400 000 years (Ince 2007, page 33).

The possibility of drastic changes in the spin axis of a planet is demonstrated by Mars. The present tilt of Mars of about 25 degrees is similar to the tilt of the Earth at just over 23 degrees which results in the seasonal variations being similar on the two planets. However, Taylor (1998, page 123) writes that this similarity is coincidence; the spin axis of Mars changes every few million years tilting between zero and 60 degrees. Mars may have more extreme variations in its climate than the Earth and could have been far warmer a few tens of thousands of years ago (Ince 2007, page 33). The change in axis would cause the polar ice caps on Mars to melt as the angle changes.

The Milankovitch cycles are regular cycles affecting the Earth. The Earth's changes in tilt on its axis, wobble which causes the precession of the equinoxes and eccentricity of its orbit around the Sun are motions which occur on a cycle of thousands of years and which cause the climate to go through a cycle of variations. The Milankovitch Hypothesis gives an explanation for the fluctuations of ice advance and retreat within an ice age. The hypothesis does not explain why ice ages occur in the first place.

The Arctic Circle and Antarctica
There is abundant evidence that in the era preceding our own era –the Mesozoic –both Antarctica and the Arctic Circle had a mild climate. Especially during the Cretaceous period both these regions were covered by huge tracts of forest.

During the Mesozoic era between 245 and 66 million years ago Antarctica was densely forested with conifers and ferns, and the angiosperm southern beech spread during the Cretaceous period. These forests were inhabited by amphibians and reptiles. Jurassic-type dinosaur fossils have been found on Antarctica indicating a milder climate. The Antarctic continent was surrounded by warm seas inhabited by ammonites until the Late Cretaceous or early Tertiary period.

Antarctica was probably connected to Australasia and South America by land bridges. These strips of land provided a migration route to animals such as marsupials between Australasia and South America between 150 and 70 million years ago.

Antarctica began to freeze in the Cenozoic era. An ice sheet had formed in Antarctica 35 million years ago, and became more established 15 million years ago. The ice sheet over Antarctica now rises to 3000 metres above sea level. Antarctica now has an average temperature of −49° C.

Erosion of the coastal areas severed Antarctica from the other continents; peninsulas and islands either side of Drake Strait hint at the former connection of South America to Antarctica. Thus, migration routes were cut off and Antarctica is now completely surrounded by cold seas. Around the edges of the Antarctic continent the ice sheets float on the sea and are called ice shelves. These advance and retreat in winter and summer, and give rise to icebergs.

The Arctic was a landmass during the Palaeozoic era. It had thick sediments folded into mountains. This landmass has now eroded away leaving the Arctic region as sea. Now only eroded islands remain with Greenland being the largest island.

During the Mesozoic era the Arctic may have provided a land route for the migration of dinosaurs from Europe to North America during the Jurassic period and from China and Mongolia to North America during the Cretaceous period. The continental shelf which at one time would have been land is 1770 kilometres wide off Eurasia.

During the Tertiary period of the Cenozoic era the Arctic saw cooler conditions. The Pliocene epoch 2.5 million years ago marks the beginning of the northern hemisphere glaciation. During the

Pleistocene epoch of the Quaternary period (which started 1.6 million years ago) continental ice sheets advanced and glaciers became widespread in northern latitudes. They receded again in the Holocene epoch 10 000 years ago.

Today the biggest Arctic ice cap covers Greenland as a great dome of ice as much as 3350 metres thick. Other islands are also covered by ice caps. In winter the waters of the Arctic Ocean actually freeze, although in summer there is some melting and areas of open water reappear. However, much of the Arctic Ocean is covered by permanent pack ice in slowly drifting ice floes.

The lands circling the Arctic Ocean have treeless tundra and the ground permanently frozen with permafrost. Below this there are forests of trees where the average temperature reaches 10 °C in July. The tree line marks the division between forests and tundra.

Antarctica has 90 % of the world's ice, and Greenland has the rest. All the other glaciers in the world do not make up any significant percentage.

Rarity of ice ages
Louis Agassiz (1807-1873) born in Switzerland coined the term 'ice age'. Ice age is the notion that the Earth goes through cold phases in its global climatic fluctuations. An ice age will last a few million years with a glacial cycle consisting of ice sheets advancing and receding.

Analyses of deep sea drill cores show that oceanic waters cooled down 35 million years ago at the end of the Eocene period. It is thought that we have been in an ice age for the past 35 million years characterized by ice advances and retreats.

At the glacial maximum of the Pleistocene 18 000 years ago ice sheets reached their greatest extent. At this time glaciers reached as far south as the Great Lakes in North America and covered northern Europe as far as Switzerland. Temperatures were between 5 and 10° C lower than today and sea level was 120 metres lower.

At present we are in an interglacial period with higher temperatures and an ice sheet covering only Greenland and some other islands in the Arctic region. The Earth's frozen areas currently represent 10 % of its surface area. At the height of the recent ice age the figure was about 32 %.

Ice ages are rare: apart from the ice age that we have been in for the last 35 million years, ice ages are documented for 280 million years ago at the end of the Carboniferous and beginning of the Permian periods lasting a few tens of millions of years. Older drift deposits suggest an ice age 450 million years ago during the Ordovician period. A Precambrian ice age occurred 600 or 650 million years ago which lasted 25 million years. Cloud (1988, pages 223-225) writes that another Precambrian ice age occurred 2300 million years ago and left evidence all over the world. Another ice age is thought to have occurred 3000 million years ago. This gives a total of 6 ice ages throughout Earth history.

Explanations for ice ages

Snowball Earth Theory
A popular contemporary theory is the Snowball Earth Theory proposed by Paul Hoffman [17]. Hoffman suggests that the Earth underwent a super ice age during the Precambrian period 600 million years ago in which the entire planet froze, even at the equator. This cataclysm would have been the end of life on Earth, but the Earth was saved by volcanic eruptions which warmed the planet up again.

The author Gabrielle Walker searches out evidence of glaciation in places that are now warm such as Australia, Namibia, South Africa and Death Valley, USA in support of the theory.

The idea of a global covering of ice was first suggested by Brian Harland of Cambridge University in 1964. The explanation given for the super ice age is positive feedback in the climate – if the climate gets cooler it snows more – snow and ice are white and so reflect solar energy back into space. The high albedo effect (albedo is the amount of light reflected away compared to the amount absorbed) causes a further drop in temperatures and more snow falls. This situation was only reversed by the greenhouse effect caused by volcanic eruptions.

---

[17] Walker, Gabrielle (2004) *Snowball Earth* Bloomsbury Publishing PLC

An alternative explanation has been suggested by Robert Berner after studying global atmospheric carbon dioxide levels and linking it to long-term climate change. Berner suggests that the ice age 600 million years ago could have been the result of global cooling caused by the massive drawdown of carbon dioxide when single-cell organisms started to extensively colonize shallow seas (Lamb & Sington 1998, page 164). The deposition of the tiny shells composed of calcium carbonate of marine plankton led to extensive accumulations of limestone at this time.

Continental Drift

The notion of ice ages is a 'climatic phase theory'; there is also an explanation for evidence found of glaciation as a 'geographic spatial theory'.

Evidence of glaciation is sometimes found in rocks in warm countries. For example, there are scratch marks made by a glacier in Carboniferous rocks 290 million years old in Kimberley, South Africa.

This anomalous evidence is explained by Continental Drift with the notion that land masses have changed their locations on the face of the globe. The idea is that landmasses that are presently located near the equator once drifted over a pole and became affected by glaciation during their time at this latitude.

Lamb & Sington (1998, page 164) write that 280 million years ago extensive ice sheets covered parts of the supercontinent of Gondwanaland comprising what are today the landmasses of the southern hemisphere. They also write,

"There is a general coincidence between Ice Ages and periods in the Earth's history when continents are clustered near the poles. For instance, during the early Ice Ages, around 600 million years ago, and also in a later period of Ice Ages, around 280 million years ago, large supercontinents were near the south pole." (Lamb & Sington 1998, page 171).

Continental Drift Theory will be discussed in greater detail in Chapter 21, section 2.

Comment

In my view, the Snowball Earth Theory is likely to be an exaggeration with the 'entire planet enveloped in ice' sensation not entirely true. Neither do I adhere to the notion of Continental Drift.

Whatever the state of global climate, it is always true that the poles are colder than the equator. This is because the Sun's rays reaching the poles hit the Earth at an oblique angle imparting less heat per given surface area. At the equator the radiation hits nearly head-on concentrating heat for a given surface area which produces a tropical climate in these regions. In this way, the angle at which solar radiation hits the Earth results in a temperature gradient from the poles to the equator; the poles are always colder than the equator; ice sheets form mainly at the poles.

The occurrence of glaciations appears to be affected by a combination of many different factors which include carbon dioxide levels in the atmosphere; ocean currents that redistribute warm water around the planet; the occurrence of landmasses at the poles since ice sheets form mainly on land; volcanic episodes launching dust into the atmosphere that causes global cooling; and Milankovitch cycles relating to the Earth's spin axis and eccentricity of the orbit affecting the seasons.

I speculate whether the six major ice ages which have occurred over Earth history at irregular intervals of between 170 and 1650 million years apart could have been provoked by capture events in the solar system. Maybe the spin axis of the Earth has veered between more upright and tilted when disturbances occurred.

An upright planet is more uniformly warm than a tilted planet; a tilted planet has seasons. A phase of greater tilt could have produced more extreme seasons at the poles and led the Earth into an ice age. A more upright Earth would be less seasonal with a more even distribution of temperature. Under these conditions the poles may experience a temperate climate allowing them to support temperate vegetation.

Seasonality caused by the tilt of the Earth could have gone through a series of changes through geological time. There is also evidence that the Earth has been globally much warmer throughout most of Earth history than it is now. The Oligocene epoch 35 million years ago saw a global cooling with the spread of grasslands. Grasslands can withstand seasonal periods of cold and dry weather by springing up afresh when winter or the dry season is past. During the Oligocene, ice sheets spread over Antarctica and later during the Pliocene epoch ice sheets spread over the Arctic region.

Among the complex reasons for ice accumulation, greater tilt in the Earth's axis may be a significant ultimate cause. As far as extinctions are concerned, if the Earth's tilt changed, the seasons would change in their degree of severity. This would cause a shifting of climatic zones and challenge organisms to adapt to new conditions. It would lead to the dying out of species and the colonization of new species as the climatic zones shifted.

4. Episodes of volcanism

Episodes of volcanism that have occurred in various parts of the Earth at various times may have been provoked by decompression of the mantle. Decompression of the mantle may have been caused by events in the solar system. This is the subject of this section and section 5.

Volcanoes

Volcanoes are formed by eruptions of molten magma emanating from the Earth's mantle. The hot, liquid rock from the Earth's interior is known as magma, and when it cools down and solidifies on the Earth's surface it is known as lava. Volcanism builds up the Earth's crust with basalt rocks by transporting silicate minerals from the mantle to the surface.

Volcanoes may erupt violently, ejecting pieces of rock material, ash and dust, and gases into the atmosphere. Volcanic gases mostly consist of carbon dioxide, but also sulphur dioxide, and water vapour. The erupting type of volcano builds up a cone of lava and ashes around a central crater.

Volcanic eruptions under sea tend to ooze out magma rather than erupt. This is due to the greater atmospheric pressure deep under water. Volcanoes under the oceans form fissures of molten magma that form oceanic ridges. A typical lava form produced underwater by oozing magma is pillow lava. Pillow lava consists of rounded mounds of lava shaped like pillows.

Molten basalt magma has a temperature of 1200-1100 °C. Pyroclastic flows consisting of super-heated ash and gases erupt from volcanoes at between 100 and 700 °C. Geysers found in volcanic areas let off steam and bubble out hot water at 300 °C.

Magma domes
The Earth's mantle is formed of solid rock minerals but parts of the mantle can melt forming subterranean chambers of molten magma within the upper mantle. The chambers of magma are periodically emptied via conduits to the surface resulting in the eruption of volcanoes.

One of the most volcanically active regions on Earth is Yellow Stone Park in Wyoming USA. Yellow Stone has a terrain of broad volcanic plateaus on average 2440 metres high and the Yellow Stone Caldera which is a super volcano. The volcanic features of the area which include hot springs and geysers are the result of Yellow Stone Park sitting over a giant magma chamber within the Earth's upper mantle.

Archipelagos of volcanic islands are often found above plumes of magma that have pushed up the oceanic crust into broad domes often over 1000 km across. For example, the volcanic islands of Hawaii are formed above a dome covering a wide area that makes the ocean 1.5 km shallower than the surrounding deep ocean. The Hawaii Island volcanoes are found in the middle of what is known as the Pacific Plate.

(Note that a volcanic dome is not a lava dome. A lava dome is a rock formation composed of solidified lava from a volcano).

Decompression
If the Earth's mantle is solid rock —it rings like a bell during earthquakes —how and why does it form chambers of molten liquid rock?

The reason why the mantle is formed of solid rock is that despite having a temperature at which silicate minerals would normally melt, it remains solid due to being under very high pressure. The pressure from the overlying layers of rock in the mantle and crust keeps the mantle solid. If this pressure is released, the minerals of the mantle melt, become magma and expand upwards. Thus, chambers of magma in the mantle are associated with decompression of the upper mantle.

Volcanoes have their root zone in the upper mantle at a depth of between 70 and 200 km below the surface. This represents only the outer layer of the mantle since the total thickness of the mantle is 2900 km. Magma has a temperature of close to 1200 $^{\circ}$C when it is located a few kilometres below the surface. Therefore, basalt magmas are formed by decompression melting of the mantle.

Volcanic hot spots
There are many so-called volcanic hot spots around the world dating from different geological eras. The volcanism has been so intense in many hot spots that volcanoes have given rise to flood basalts forming high plateau regions. The volcanic episodes can be dated from the composition of the basalts:

- The Siberian Traps date from 250 million years ago.

- Parana where the Iguassu Falls are located in South America dates from 120 million years ago.

- The Deccan Traps in India date from 65 million years ago and erupted for over a million years.

- The Ethiopia flood basalts date from 25 million years ago.

- The Columbia River Plateau Basalts in North America date from 17 million years ago (Lamb & Sington 1998, page 108).

Hot spots on the Earth's surface for volcanic activity have been explained by Plume Theory or the Hot Spot Hypothesis as a back-up to Plate Tectonics. Plate Tectonics and Plume Theory will be discussed in section 4 of Chapter 21. (Plume Theory asserts that plumes of hot magma rise up through the mantle from the Earth's core-mantle boundary and burst through the crust as volcanoes. The problem with the theory is that seismic studies show that the plumes of magma do not come from the lower mantle, but much more superficially from the upper mantle).

The link between hot spot regions and episodes of volcanism that can be dated suggests a possible cause linked to solar system events. The location of hot spots in the middle of so-called tectonic plates rules out their explanation with Plate Tectonics Theory.

Volcanism and solar system disturbances

Decompression of the mantle leading to an episode of volcanism in a region on Earth's surface could be produced by the initial capture event or by interaction with other captured bodies.

Capture

In Chapter 17 I wrote about capture into the solar system of Earth, other planets and the satellites of the Jovian planets. I suggested that a capture event would be marked by cratering and volcanism affecting the captured planet.

The Earth underwent an intense thermal period of volcanic activity 3.8 thousand million years ago. This would mark the event of capture of the Earth into the solar system. Other terrestrial planets also show patterns of volcanism and cratering.

Mars shows the pattern of having a low terrain with many craters in the southern hemisphere and many large volcanoes in the northern hemisphere. Mars has a huge bulge on one side of the planet 10 km high and 8000 km across called Tharsis. The Tharsis region covers a quarter of the planet. It has been built up by volcanic activity that has continued over 2000 million years.

Venus also shows a pattern of low land and cratering on one side of the planet, and high land and volcanism on the opposing side. The southern part of Venus has relatively flat rolling terrain with some craters, while the northern region consists of upland plateaus with no craters and large numbers of volcanoes.

If a terrestrial planet was captured into the solar system, the leading edge would be affected by impacts making craters while the trailing edge may be affected by volcanism. The pattern could be explained by compression of the mantle on the leading edge and decompression of the mantle on the trailing edge.

The Earth has land masses almost entirely missing on one side – where the Pacific Ocean covers almost a third of the surface of the globe.

The Earth could have been captured into the solar system with the Pacific Ocean representing the leading edge. Many craters would have been excavated on this side of the planet by collision with debris in the solar system and these large craters became the basins of the Pacific Ocean. The opposite side of the planet where Africa is now located would have experienced decompression of the mantle and an early episode of volcanism. The basalts from these volcanic eruptions, now metamorphosed into serpentine rock became the core of the African continent.

The case against this scenario is that it only works if the geographic poles were located over the Pacific Ocean and Africa at the time of capture. If these regions were on the equator as they are at present, the leading edge would be a revolving edge, not a region.

Disturbances
As noted above, Earth history has been marked by various episodes of volcanism producing 'hot spots' at various locations around the planet.

Apart from a planet's own capture event, the capture of other planets could cause disturbances in other already resident planets. The different episodes of volcanism on planet Earth producing 'hot spots' on its surface which are dated to different geological periods, could be explained in terms of the capture of other planets into the inner solar system or of bodies passing close by the Earth before being engulfed by the Sun.

The gravitational interaction between the Earth and a planet or body passing close by will be examined in the next section.

5. Gravitational waves

Gravitational waves are produced by moving masses in a similar way to electromagnetic waves being produced by moving electrical charges. Gravitational waves, also known as gravitational radiation are extremely weak. No instrument has been able to detect gravitational waves yet, but there is confidence that such waves exist from studies of binary star systems.

Quadrupole moment

According to Einstein's General Relativity Theory, the amount of gravitational waves that a mass can emit depends on its quadrupole moment. The quadrupole moment is related to the shape of the mass. A perfectly spherical mass does not radiate gravitational waves whereas an elongated shaped mass does. "An American football has a large quadrupole moment, but a soccer ball has none." (Gribbon & Rees 1991, page 203).

Gravitational waves have a distinctive effect on spacetime as they pass through it. Gribbon & Rees suggest that the quadrupole radiation effect can be visualized by thinking of a flexible circular ring.

"When a gravity wave passes it, the ring is stretched in one direction and squeezed in another, at right angles, simultaneously. It becomes an ellipse. Then, the pattern reverses, and what was the long axis is squeezed while what was the short axis is stretched. This pattern of alternate squeezing and stretching, in two directions at right angles to each other and out of step, is the characteristic "signature" of quadrupole gravitational radiation. It is not just the ring that is actually being stretched and squeezed, but the fabric of space itself." (Gribbon & Rees 1991, page 203-204).

It can be observed that the Earth is not perfectly spherical –it has a rotational equatorial bulge. When measuring the gravitational potential at the surface of the Earth, the largest variation in potential around the Earth is due to the equatorial bulge. There is an increase in the value of gravity on the surface of the Earth from the equator to the poles (Encyclopedia Britannica: gravitation).

Measurements of the gravitational potential also vary depending on the positions of the Sun and the Moon. Thus, the elliptical orbit of the Earth affects the measured strength of gravity. In my view, the shape of the Earth and its orbit indicate that the Earth could have a quadrupole moment.

19.1
GRAVITATIONAL WAVES AND QUADRUPOLE MOMENT

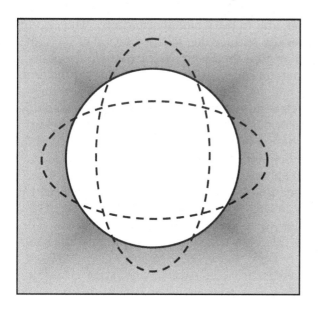

When gravitational waves pass through a sphere, the long and short axis are stretched and squeezed alternately. (This illustration was redrawn from Gribbon & Rees 1991, page 204).

Tides
A tide is, according to Encyclopedia Britannica "any of the cyclic deformations of one astronomical body caused by the gravitational forces exerted by others."

Ocean tides

We usually think of tides as the changes in sea level of the oceans on Earth caused by the gravitational pull of the Moon and the Sun. At the surface of the Earth the gravitational force of the Moon is around 2.2 times greater than that of the Sun. Tides are produced by the Moon due to variations in its gravitational field over the surface of the Earth compared to its strength at the centre of the Earth.

The effect of the gravitational pull of the Moon is that water tends to accumulate on the parts of the surface of the Earth directly toward and directly opposite the Moon, and to be depleted elsewhere. The regions of accumulation move over the surface due to the rotation of the Earth and the orbit of the Moon around the Earth. The two high tides and two low tides produced each day are modified by the effect of the Sun.

The height of ocean tides depends on whether the lunar and solar tides are working together or in opposition. The bulges raised by the Sun and Moon can either be aligned or at right angles. If they are aligned, the result is a very high tide followed by a very low one. This is a 'spring tide' – spring as in spring back and forth a long way, not the season. Coincidentally, the highest solar tide is in January because the Earth is closest to the Sun at this time of year. When the two bulges are at right angles, the tides are lower and are called 'neap tides'. The actual height of water of a tide is in large part determined by the shape of inlets of sea.

Earth tides

There are also gravitational effects on the interior of the Earth. An Earth tide is an elastic deformation of the Earth, rather than a flow. Observations of Earth tides contribute to knowledge of the internal structure of the Earth.

The Moon and Sun set up two bulges in the oceans, the atmosphere and the Earth itself, one pointing directly towards them and the other directly away from them. With Earth tides the Moon raises a tide about one metre high at the Earth's equator while the Sun raises a tide about half a metre high. It is possible that these Earth tides provoke volcanic eruptions and earthquakes. Ince (2007, page 36) writes that when the lunar and solar earth tides coincide to form one major tide, a significant earthquake or volcanic eruption is more likely.

The Encyclopedia Britannica article on tide finishes with the observation that tidal processes can also occur on other members of the solar system. It has, for example, been suggested that the volcanic activity on Io, one of Jupiter's satellites, could be caused by internal heating by frictional resistance to tidal deformation.

On the basis of the above information on quadrupole moments and tides, I suggest that if other planets flew by the Earth into the inner solar system or into the Sun to be engulfed, they would form a fleeting binary system with the orbit of the Earth. The effect of a close fly-by would be the production of gravitational waves that would alternately squeeze and flex the body of the Earth in a quadrupole motion. If the gravitational waves were strong and violent enough, the Earth's mantle would be significantly compressed and decompressed on the side towards and the side away from the approaching body. Decompression of the mantle in this way could lead to an episode of volcanism in particular parts of the Earth.

Due to gravitational interactions between bodies and the production of gravitational waves, episodes of volcanic activity on Earth may mark the capture of the Earth into the solar system, the subsequent capture events of other planets into the inner solar system and the capture of large bodies engulfed by the Sun.

6. Tidal waves

In this chapter I have noted that the capture of planets into the inner solar system, the capture of bodies that plunged into the Sun and the collision of satellites with the giant planets in the outer solar system would produce disturbances in the solar system with major effects on Earth. These effects may include changes in climatic zones with the extreme being the onset of an ice age and episodes of volcanism lasting millions of years affecting various parts of the Earth. There is another more immediate effect which would devastate the Earth and contribute to mass extinction – it is the production of tidal waves of gigantic proportion.

The gravitational pull on the oceans of the Earth of a passing body would produce tidal waves or tsunamis of such force that they would wash over the surface of low-lying landmasses sweeping animals into collective graves beneath sediments that would later become fossil beds.

Mass extinctions have simultaneously affected both animals and plants on land, and all types of organisms living in the surrounding seas. At each crisis reef-building organisms living in tropical seas have been severely affected.

If massive tidal waves had swept the planet, the land would have been flooded at the same time that disturbances to the seabed occurred through the deposition of sediments in coastal waters. Huge quantities of sediments would have been shifted by these sudden events. Tidal waves would thus explain the concurrent devastation of life on land and in the oceans during mass extinctions.

Fossil bed formation
Many seem to have assumed, with the concept of gradualism that fossil beds were formed over the course of millions of years very slowly with dead animals gradually becoming encased in sediments. Gradualism is the belief that new species evolve gradually over time.

In fact, few animals die of old age in the wild. The infirm are picked off by predators and consumed. The ones that do die of disease do not leave carcasses for the fossil record since their flesh and bones are scavenged and scattered. Decomposition effaces the last remains of animals and plants under normal circumstances within ecosystems.

It is sudden burial beneath layers of sediment that precludes the normal food chain of predation and scavenging and preserves animals and plants from the microbial processes of decomposition which occur in the presence of an aerobic atmosphere.

In my view, fossils are preserved within sedimentary strata firstly due to recurrent localized disasters such as the flooding of rivers covering flood plains with silt or erupting volcanoes covering some areas with ash. Secondly and more importantly, would be the effects of tidal waves or tsunamis washing over land during large global catastrophes. Rich fossil beds would be formed by the sudden burial of many types of living animals and plants swept by cascading waters to their grave beneath millions of tons of sediments.

Evidence for the megadisaster formation of fossil beds is that many fossil beds are tens of metres deep. The formation of fossil beds in this sudden way would also account for the excellent state of preservation of many whole specimens.

As far as the appearance of new species in the fossil record is concerned, the new species would appear in new higher strata that may have been laid down a long interval of time after the underlying strata.

Many famous fossil beds marking the end of geological periods could have been produced by megadisasters involving water-borne sediments.

As an illustration of deep fossil beds I will give three examples of fossil beds containing dinosaur bones:

Dinosaur National Monument
At Dinosaur National Monument in Utah and Colorado, USA an area that is part of the Rocky Mountains has been cut by rivers to form deep canyons. Above the Green River there is a rock face 65 metres high in which sediments containing the fossilized bones of many dinosaurs has been exposed through the processes of erosion. The first excavations of 1909 to 1923 yielded 350 tons of dinosaur bones. These included many reasonably whole skeletons that could be mounted.

The dinosaurs discovered at Dinosaur National Monument lived during the Jurassic about 150 million years ago. They include the bones of herds of plant-eating giant sauropods such as *Apatosaurus*, *Camarasaurus*, *Diplodocus* and *Barosaurus*. The plated *Stegosaurus*, another plant-eater was very common. *Camptosaurus* would have browsed the leaves of ginkgo trees. *Dryosaurus* was a smaller fast-moving plant-eater. The plant-eating dinosaurs would have provided food for the carnivores including the giant *Allosaurus*, horned *Ceratosaurus* and small *Ornitholestes*.

Young dinosaurs such as metre-long *Allosaurus* died along with the adults. The deposit also preserves the remains of crocodiles, turtles and frogs. At the time of the Jurassic, the area had lush vegetation watered by rivers which left sand-bars and mud that preserved the footprints of dinosaurs. It has been suggested that the dinosaur bones accumulated by frequent flooding of these Jurassic rivers, but I propose that the dinosaur graveyard was formed by a more cataclysmic type of flooding.

Dinosaur Provincial Park
Dinosaur Provincial Park in Alberta, Canada is a barren 'badlands' area. The badlands scenery has been produced by the Red Deer River and rain and snow eroding ancient sediments into steep-sided rock formations. Erosion of the sediments has exposed the bones of dinosaurs that lived 75 million years ago during the Cretaceous.

The fossils include the skeletons of a whole herd of *Centrosaurus* preserved in one fossil bed of bones. The bones are of both young and old animals, and some are broken and trampled as if they had died in a stampede.

The bones of *Hadrosaurus* also known as 'duck-bill dinosaurs' are common in the deposit. Another plant-eater was the armoured *Euoplocephalus*. These dinosaurs would have been preyed upon by the ferocious tyrannosaurid *Albertosaurus*.

I believe that the 'badlands' provides an example of a landscape formed by the rapid deposition of huge quantities of sediments, and that the dinosaurs died surprised by sudden catastrophe.

Trossingen
The site at Trossingen, Germany is also suggestive of sudden burial by sediments brought by the flooding of a megadisaster. At this site masses of complete skeletons and individual bones of *Plateosaurus* have been found. They appear to represent a migrating herd of plant-eating prosauropod dinosaurs.

Fossil fuels
Fossil fuels are beds of organic plankton and plant remains buried deep underground. Fossil fuels may be another testimony that cataclysmic events have taken place on Earth. Fossil fuels include crude oil also known as petroleum, natural gas, coal and bitumen or asphalt also known as tar. These organic deposits are found encapsulated in sedimentary rocks.

I envisage the deposition of huge quantities of sediments upon living ecosystems by giant tidal waves sweeping the surface of the Earth when solar system disturbances occurred. In this way organic matter would have been buried and preserved, and later transformed into hydrocarbon compounds.

Oil
Oil is formed from unicellular organisms that lived during the Precambrian. They include photosynthetic cyanobacteria and algae such as diatoms that lived a planktonic existence as well as protozoan foraminifera. Oil fields are buried sedimentary basins.

I envisage that the sedimentary basins of oil fields represented ancient pools formed in craters in which cyanobacteria and algae flourished, building up large accumulations of organic matter at the bottom of the pools. Cataclysmic events caused these basins to be filled up by further sediments, trapping and burying the accumulated organic matter. The sealing-in of organic matter by sedimentary strata allowed the processes of petroleum formation to take place.

In the first stage of the formation of petroleum, anaerobic bacteria transformed the organic matter of photosynthetic bacteria and protists into kerogen giving off methane gas. With deeper burial by continuing sedimentation and with increasing temperatures kerogen is transformed into crude oil by thermal degradation and cracking (the process by which heavy hydrocarbon molecules are broken up into lighter molecules). Crude oil consists of many different complex hydrocarbon molecules.

Petroleum migrates upwards into porous sedimentary reservoir rocks where it is trapped by impervious layers of rock above or seeps to the surface of the Earth.

Worldwide there are 600 sedimentary basins with giant oil fields and another 240 basins with smaller oil fields. The greatest concentration of supergiant oil fields is in the Persian Gulf where they number twenty, while in the rest of the world adjacent supergiant oil fields never exceed two. There are supergiant oil fields in the United States, Mexico, Venezuela, Russia, Libya, Algeria and China. There are many other oil fields that are small in size.

I see the distribution of oil fields around the world as an indication of the location of pools of water inhabited by thriving photosynthetic unicellular life during the Precambrian. The pools may have been the size of lakes or even of small seas, and some may have been formed in craters produced during the capture event.

The location of oil fields may be an indication of which regions had a warm climate during the Precambrian allowing photosynthetic organisms to flourish. Oil fields in Antarctica indicate that this now frozen continent had a warm climate conducive to blooms of cyanobacteria and algae for at least some periods during the Precambrian.

Coal

There is a type of Precambrian coal called anthracite formed from algae, diatoms and fungi living in water 2500 million years ago. Most other coal is formed from the remains of trees that lived during the Carboniferous, Permian and Cretaceous periods.

Peat deposition is the first step in the formation of coal. Peat bogs are found today in wet temperate climates. Normally plant matter from dead plants decomposes into its basic components, many of which are gaseous; this prevents an accumulation of organic matter to a large extent (although organic matter contributes to the formation of soil). However, where plants grow in bogs, when they die the vegetable matter falls into the boggy waters where there is insufficient oxygen for bacteria to decompose it fully. The result is peat which is brown and spongy, or black and compact depending on the degree of decomposition. Peat is burned as a fuel in considerable quantities in Ireland, Russia, Sweden, Germany and Denmark. There are vast beds of peat in Europe, North America and northern Asia.

Peat is converted into coal when it is buried beneath layers of sediment. Under pressure from the over-lying sedimentary strata, the peat which at first contained over 90 % water dries out and hardens into low-grade coal called lignite. After more time and greater pressure it becomes bituminous coal and finally high-grade anthracite coal. Coal gives rise to viscous bitumen and asphalt or tar.

The first continents on Earth were low-lying landmasses with a vegetation of swamp forests. The forests were dominated by the first tree-sized vascular plants of club mosses, horsetails and seed-ferns. These forests grew up during the middle Devonian and covered the land during the Carboniferous and early Permian periods.

These vast swamps with their luxuriant primitive trees accumulated plant matter that at certain points in time were buried under layers of sediment. It has been suggested that the sediment was brought by periodic incursions of the sea on land. I suggest that these incursions of the sea could have been caused by catastrophic events.

We now rediscover the Carboniferous swamp forests as beds of coal. Deep beds of coal of Carboniferous age are found in Alaska, Canada (Nova Scotia), North America (Pennsylvania), the British Isles, Central Europe and Southern Russia. Coal of Permian age is found in Siberia, East Asia and Antarctica.

There is also coal of Cretaceous age formed from angiosperm (seed-bearing) trees. Cretaceous coal is found in western North America, Alaska, southern France, Central Europe, Japan and Indonesia.

Therefore, the happy occurrence (for the economy) of fossil fuels such as oil wells and coal seams may, in fact, denote the occurrence of catastrophes on Earth accompanied by mass devastation of the Earth's Precambrian cyanobacterial mats, planktonic algal ocean populations and forests covering the land in former eras.

Conclusion

If one adhered to a uniformitarian view of geology – the idea that geological processes have always gone on at the same steady rate as we observe now, but for a very long time – then one would oppose the view that large scale events had occurred in the solar system because if they had, catastrophes would have occurred on Earth. But when we look at the geological record, we find evidence for catastrophes on a massive scale – cataclysmic disasters – episodes when much of life on Earth went extinct. This is the subject of the Mass Extinctions Hypothesis.

Possible changes to Earth's axis of rotation would affect the degree of seasonality in temperate regions of the Earth. More marked seasons could initiate ice ages with the ebb and flow of glacial maxima and interglacials. Ice ages affect global climate and the distribution of ecosystems over the globe. The onset of an ice age would cause a shifting of ecosystems, rather than sudden extinction.

Earth tides and even gravitational waves may cause decompression in parts of Earth's mantle leading to episodes of volcanism affecting different regions at different times. A region affected by flood basalts would undergo the destruction of its ecosystem, but episodes of volcanism are unlikely to be linked to mass extinction on a global scale.

The effects of gravitational attraction on the oceans of Earth producing giant tidal waves that washed over landmasses is more likely to be the main factor in mass extinctions with the concurrent formation of deep fossil-bearing strata. Giant tidal waves would account for the burial animals, plants and whole forests on land. At the same time, the washing out to sea of sediments would also bury marine invertebrates living on the seabed, coastal fish and marine reptiles, as well as submerging coral reefs close to the shore line.

The tsunami effect could be produced by gravitational interaction with large bodies in fly-bys passing close to the Earth. There would be a gravitational pull on the waters of the oceans on one side of the Earth followed by a sudden back flow. The large bodies would have disappeared into the Sun or have taken up orbit in the inner solar system. Another solar system event could be shock waves produced by the collision of a large body with a planet.

At the time of the asteroid belt collision which sent not only shock waves, but also asteroids spinning in all directions, several or even many asteroids could have hit the Earth within a short period of time landing on land and in the oceans. Asteroids striking the oceans would produce giant tsunamis.

Killer tidal waves fit the bill for the mass extinction of life on land beneath layers of sediment at the same time as disrupting the seabed and the extinction of life in the sea.

Mass extinctions of life on Earth are testified to by deep fossil beds and the deep burial of organic matter that has been transformed into fossil fuels.

The idea that the foundations of the Earth are not firm after all is maybe not a very welcome idea. Earth history testifies to upheavals, the scale of which indicates that the causes are likely to have come from disturbances produced outside the Earth itself.

Giant tidal waves washing over land would devastate low-lying land. The main habitat and food source of the dinosaurs of the Cretaceous may have been swept away. Any large dinosaurs not swept away would have been left with insufficient vegetation to feed on. But the disaster would not have wiped out the plants and animals of high mountain tops. Some small marsupials and primitive placental mammals clinging to some flowering trees on high mountains did survive. This empty world was their new opportunity.

Bibliography

Abell, George (1969) *Exploration of the Universe* Holt, Rinhart & Winston

Canby, Thomas Y. (1993) Bacteria: Teaching Old Bugs New Tricks. *National Geographic*, August 1993, pages 36-60.

Cloud, Preston E. (1988) *Oasis in Space: Earth History from the Beginning* W.W. Norton & Company, New York

Encyclopedia Britannica 2011 Standard Edition: Antarctic; Arctic; Arctic Circle; Bitumen; Coal; Cretaceous Period: Mass extinction; Dinosaur; Gravitation; Iron; Magnetic field; Magnetic permeability; Magnetism; Oil shale; Peat; Petroleum; Permian Period: extinctions; Tide; Volcanism; Volcano

Gould, Stephen Jay (ed.) (1993) *The Book of Life.* Ebury Hutchinson

Gribbin, John & Martin Rees (1991) *Cosmic Coincidences: Dark Matter, Mankind and Anthropic Cosmology* Black Swan

Hunten, Donald M. (1993) Atmospheric Evolution of the Terrestrial Planets. *Science* Vol. 259, page 915.

Ince, Martin (2007) *The Rough Guide to the Earth* Rough Guides Penguin Group

Lamb, Simon & David Sington (1998) *Earth Story: The Shaping of Our World*
BBC Books

Taylor, Stuart Ross (1998) *Destiny or Chance: Our Solar System and its Place in the Cosmos* Cambridge University Press

The Hutchinson Encyclopedia (2004) Hodder Arnold: Ice age, Milankovitch.

Vogel, Gretchen (1999) Expanding the Habitable Zone. *Science* Vol. 286, pages 70-71.

Wilson, E.O (ed.) (1988) *Biodiversity* National Academy Press

Zeilik, Michael (2002) *Astronomy: The Evolving Universe* 9th edition Cambridge University Press

CHAPTER 20

THE ORGANIC ORIGIN OF ROCK STRATA

Introduction

In this chapter I am going to pursue the theme that life created the environmental conditions on Earth that favour life. In Chapter 10 of Part I and Chapter 18 of Part II I discussed the formation of water as a by-product of metabolic processes. I proposed that the different forms of life, and especially microbial life, have produced water forming the oceans that now cover our planet.

In this chapter I will focus attention on rocks, and especially those forming continents. Continental crust is a unique feature of the Earth, compared to other planets in the solar system. Although it is not an accepted view, I propose that it is life – by its metabolic activities, digestive activities and the secretion of protective structures around single cells and around primitive colonial animals – that have transformed Earth's minerals and largely contributed to the building of landmasses.

In this chapter I wish to challenge the simple division of the world into life and non-life categories. The early scientists of the 17th and 18th centuries classified the natural world into three broad categories: animal, plant and mineral. The mineral category represented non-life – a non-living substrate upon which organisms live.

The mineral category has generally been kept disconnected and separated from any association with life forms. Insights into the processes of life have not been used to throw light upon the transformation of minerals, and the deposition of sedimentary rocks except in the cases of rocks that contain obvious fossils such as coralline limestone. Until now, geology has been a science that only admitted the operation of physical processes.

The traditional classification of the world into life and non-life categories has not really changed and scientific disciplines run along the lines of a division between physical processes and the activities of life. For anyone who looks out on the world, the division between non-life and life seems both simple and obvious – rocks and minerals belong to the former category; however, science is not always about the obvious. I believe that the uniqueness of our planet is intimately connected to the processes of life.

1. Igneous, sedimentary and metamorphic rocks

Silicon plays a structural role in rocks as carbon does in cells. The main component of Earth's crust is silicate minerals (95 %). Silicon prefers to bond with oxygen, and silica or silicon dioxide (SiO_2) forms sheets or frameworks. The silicate groups consist of one silicon atom surrounded by four oxygen atoms. The silicon atom may be replaced by a metal atom in silicates.

Although most of the rocks on Earth are composed of oxygen and silicon, there are also rocks that contain calcium or magnesium instead of silicon. The mineral calcite and the rock limestone are composed of calcium carbonate ($CaCO_3$). Calcium is also found in the structural tissues of plants and animals.

A mineral is a solid chemical compound with a characteristic chemical composition and a highly ordered crystalline structure. A rock is an aggregate of various minerals with no fixed chemical composition.

Some of the oldest rocks discovered on Earth are rocks in the Isua hills of Greenland dated to 3.75 thousand million years old by the rubidium-strontium dating method. They consist of volcanic lavas that must have erupted under water and fine-grained sedimentary rocks. The Barberton rocks of South Africa and Swaziland were deposited between 3.4 and 3.2 thousand million years ago. Rocks in Minnesota have been dated to 2.7 thousand million years. The greenstone belts are heavily metamorphosed basalts of Precambrian age. The greenstones of Western Australia were metamorphosed 2.7 thousand million years ago, but originated over 3 thousand million years ago. In other places rocks are less than 2.7 thousand million years old.

According to Shaw (2006, page 201) the oldest minerals on Earth dated by radiological dating are zircons from Australia which are 4.2

thousand million years old, and the oldest rocks on Earth are the Acasta gneisses from Canada dated to 4.08 thousand million years.

In the current classification of rocks there are three main types:

- Igneous rocks are those that are produced by volcanic action or intense heat beneath the Earth's surface. Igneous rocks come from magma. The main examples are basalt and granite.

- Sedimentary rocks are formed by deposition of mineral or organic material, or precipitation of minerals from solution followed by consolidation. Sedimentary rocks are formed at the surface of the Earth often in water and build up in layers. Examples include sandstone, shale, limestone, conglomerate and breccias.

- Metamorphic rocks are igneous and sedimentary rocks that have been altered by heat and pressure below the Earth's surface. An example of a sedimentary rock transformed into a metamorphic rock is the transformation of limestone into marble.

2. Early geologists

Abraham Werner
One of the first people to form a theory of the origin of rocks was Abraham Gottlob Werner in Germany in the late 18th century. His work in the mining industry led him to observe that in areas such as the Bohemian Massif the sequence of rocks started at the bottom with crystalline rocks such as granite, gneiss and schist, and above them were a succession of layered rocks composed of sandstones and shale.

Werner imagined the world covered with water like a great flood. These waters contained all the rocks of the Earth held in suspension or dissolved as chemicals. As the waters subsided the particles held in the water precipitated or settled out to form layers of rock. Werner hypothesized that at first the chemicals dissolved in the water crystallized out on the ocean floor forming crystalline rocks such as granites, and later the suspended particles formed beds of sandstones and shales. Thus, Werner made the first attempt at a description of sedimentary processes forming rocks.

James Hutton

James Hutton from Scotland developed the theory of geology further in the 1780s by recognizing the importance of heat supplied by volcanoes in forming the Earth's crust. Hutton's observations of granite showed him that it had once been molten rock which had intruded into older sedimentary rocks of limestone and shale. Hutton hypothesized that granite is a type of lava that comes via volcanoes from the Earth's interior, but does not reach the surface before solidifying into rock. Thus, Hutton added a category of rock known as igneous rock.

Hutton believed that sedimentary rocks are made up of the fragments of older rocks that have been eroded, carried away by rivers and redeposited in water. Hutton imagined an ancient landscape eroded by rivers, shifting fragments of bedrock as sediment down to the sea. The material carried by rivers accumulated at the bottom of the sea to form shales, silts and sands that eventually became compacted into horizontal layers of rock. Later, movements within the earth lifted the rock layers out of the sea and in the process turned some of the horizontal layers vertical, while contorting and folding other layers. Thus, land rose up from the sea.

These early theories of geology did not have a timescale. Hutton believed that God had set up the Earth as a perpetual motion machine with no beginning and no end. Thus, Hutton believed that rocks were of infinite age, that is to say, they had existed forever. These ideas changed when William Thomson, Lord Kelvin a physicist from Ireland developed theoretical ideas on the conservation of energy in 1851 that led to the formulation of the first and second laws of thermodynamics. Kelvin showed that the Earth must be losing heat and attempted to measure the age of the Earth by heat loss. This brought in the notion that the Earth and its rocks were of a finite and measurable age.

Geology from its inception has been a science of physical processes. It deals with interactions between water and rock materials in mechanical weathering, chemical weathering, and the erosion and transportation action of glaciers and rivers. It deals with the transformations wrought by heat and pressure in lithification. It is a science of crystals with hypotheses relating to crystallization in magma to account for the composition of igneous rocks.

The idea that rocks such as limestone may be formed by minerals made by cells; that weathering may occur by acids produced by life forms such as fungi and bacteria growing on rocks; and that subsurface bacteria may be producing fossil fuels – appears to be an unwelcome intrusion into a subject that has always been the domain of physical processes alone. A mineral is still defined as a naturally occurring solid formed through geological processes. Traditional definitions of minerals excluded any organically derived material. However, since 1995 the International Mineralogical Association has adopted an organic class of minerals in recognized classification schemes.

I am going to take the trend of recognition of the organic origin of rocks much further in relating the organic history of Earth to the ways in which life built its own environment.

3. Rock-forming life forms

There are at least eight major groups of rock-forming unicellular and multicellular organisms. The minerals secreted by these cells are destined to provide protective structures for the animal or plant. Calcium and magnesium carbonate, and hydrated silica are the principle organically produced minerals. The sedimentary rocks of continental crust consist of vast quantities of these minerals.

Cyanobacteria

Cyanobacteria was previously known as Blue-green algae, but it is not an algae; it is a prokaryotic organism. Cyanobacteria secret calcium carbonate as well as trapping sediments in gel that serves to protect the bacteria from ultraviolet radiation. Protection from ultraviolet radiation would have been particularly important on the early Earth which lacked an ozone layer. Cyanobacterial communities form domed structures of limestone called stromatolites that appear like fields of mushrooms in shallow tropical seas. Stromatolites grow layer by layer as the microbial mat periodically creeps above a layer of sediment towards the light to continue its photosynthesis.

At one time cyanobacteria colonized much of the shallow seas of the Earth forming extensive areas of limestone stromatolites. The era of cyanobacteria started 3.5 thousand million years ago.

Radiolaria

These protist protozoans are also called actinopoda. Radiolarians live in the open sea and have siliceous shells. The shells have beautiful shapes and range in diameter from 20 μm to several centimetres in diameter. The radiolaria or actinopoda cell has fine projections called axopodia used to catch prey. The tiny prey adhere to the axopods and are engulfed by cytoplasm that transports the food to the inner parts of the cell where it is digested.

The radiolaria cell is protected by the shell formed of silica scales or spines. Some species have a cage of silica bars in repeating hexagonal patterns through which the axopods penetrate.

There is no sexual reproduction in radiolaria. They supplement heterotrophy by harbouring symbiotic yellow or green algae amongst their axopods which carry on photosynthesis.

There are silicate rocks called radiolarite cherts. An example is the chert found in the Bakany Mountains, Hungary that was formed in the former Tethys Sea.

 The Acantharian class has skeletons composed of rods of crystalline strontium sulphate ($SrSO_4$). $SrSO_4$ is secreted in their endoplasm. The mineral celestite is composed of strontium sulphate ($SrSO_4$) that could have originated from accumulations of acantharian radiolarians.

Diatoms

Diatoms are protist unicellular golden-brown and yellow-green algae. They are microscopic free-floating organisms found in both fresh and salt water. In my new classification I have put them under Chrysophyta which includes all single-cell photosynthetic algae with silica tests. The test has two valves that overlap like a Petri dish; it is called a frustule and flagella protrude from it. Diatom frustules are composed of organic materials impregnated with opaline silica ($Si(OH)_4$) that have beautiful patterns, different for each species.

Diatoms live in the ocean where they form a large part of phytoplankton, especially in cooler regions. They also live in fresh water and moist soil. Diatoms require dissolved silica in the water for growth, and they purify the water of it.

The silicoflagellates are similar to diatoms and in my classification they belong to the same group. The silicoflagellates have flagella for locomotion; plastids containing chlorophylls; and silica tests in two halves. Many species form cysts in winter that are embedded with silica and iron.

Some silicoflagellates that swim as marine plankton produce powerful toxins and form 'Red Tides'.

Diatoms and silicoflagellates have formed thick deposits with their opaline silica tests. These microorganisms precipitate silicon dioxide from its dilute aqueous solution in water to its hydrated, opal form to build their decorative shells. (Opal has the formula ($SiO_2.nH_2O$).

According to Atlas & Bartha (1981) the annual precipitation of silicon by microorganisms in the oceans has been estimated to be 10 thousand million metric tons. The shells of the dead organisms accumulate forming siliceous oozes and later "diatomaceous earth" deposits, also known as Fuller's earth.

It has been discovered that golden, motile algae and coccolithophorids are different stages of the same organism. The coccolith is a resting stage and it is composed of calcium carbonate. Since the Triassic period coccoliths have been contributing to chalk deposits ($CaCO_3$). The White Cliffs of Dover date from the Cretaceous period 120 – 80 million years ago. It is thought that at this time, continents were flooded creating shallow seas in marine shelf areas where planktonic golden algae thrived. Their coccolith shells accumulated on the sea floor eventually to become chalk.

Green and red algae
The green algae (Chlorophyta) and red algae (Rhodophyta) both precipitate calcium carbonate.

Stoneworts or Brittleworts are green algae that precipitate calcium and magnesium carbonate as a limestone covering in fresh waters.

They appear as a dense covering on the bottom of shallow ponds and are abundant in fresh and brackish waters.

Red algae are slender, branching seaweeds. Some become encrusted with calcium carbonate and are important for reef formation. The Burgess Shale dated to the Cambrian was a mud bank beside a reef constructed by calcareous algae.

There is some evidence that green algae and dinoflagellates (diatoms and silicoflagellates) have existed on Earth since a very early time. There are small organic fossils called acritarchs dating from 2.5 thousand million years which could be the fossil cysts of diatoms and other algae. The acritarchs found today represent the resting stage cysts of many different kinds of green algae (Chlorophyta) and possibly some dinoflagellates.

Acritarchs can be isolated from redeposited silicate sedimentary rocks using hydrofluoric acid. Microfossil acritarchs give an idea of the abundance of protist forms of life including algae and foraminiferans before and after the Cambrian period. The oldest acritarchs are 2.5 thousand million years old, but a thousand million years ago they started to increase in abundance, diversity and size. Some acritarchs developed spines as a possible defence against being grazed upon by other protists. The populations of acritarchs crashed 600 million years ago when an extensive ice age occurred, then proliferated during the Cambrian and crashed again at the Permian extinction.

Foraminifera
Foraminifera are marine protozoans that live on the sea floor. The spiral-shaped tests of Foraminifera consist of foreign matter such as radiolarian skeletons, sponge spicules and mineral grains bound by patches of cement-like calcium carbonate.

Foraminiferans produce cysts by sexual reproduction as a resistant resting stage in the life cycle to survive adverse climatic conditions. It is thought that some acritarchs may be the fossils of foraminiferan cysts. Acritarchs were common and varied 900 million years ago, although they appeared much earlier.

There are fossil giant foraminifera: *Lepidocyclina elephantina* has a test 1.5 cm thick; *Camerina laevegata* is a large 10 cm wide foram that lived in warm shallow waters from the Eocene to the Miocene (Margulis & Schwartz 1982, pages 128-129); and *Stannophyllum* measured 25 cm in diameter, but was only 1 mm thick – it fed on bacteria on the sea bed.

It is said that the earliest Foraminifera appeared in the Cambrian; however, the fossil acritarchs suggest that foraminifera had been around much longer. I wonder if some of the Ediacaran fauna were types of foraminifera, such as *Dickensonia* which was a flat, oval shaped animal the size of a hand.

Foraminifera give rise to limestone. In the Triassic period 230 million years ago, massive shell accumulations of foraminiferans formed thick deposits that now appear as beds of limestone in Europe, Asia and Africa. Giant foraminifera formed nummulitic limestone known as 'coin stone' with which the pyramids in Egypt were constructed. The shells of planktonic foraminifera are today settling and accumulating over much of the ocean floor as thick deposits of calcareous ooze. This is the limestone of the future.

Foraminifera such as *Stannophyllum* that fed on bacteria contained large concentrations of barite (barium sulphate) crystals called granellae in their cytoplasm. It is possible that the mineral barite ($BaSO_4$) came from this source.

Sponges

Moving now from protists to metazoans, the sponges are the most simple of multicellular animals. Their scientific name is Porifera – pore bearers. The sponge supports itself by a skeleton of needle-like spicules. Some sponges precipitate calcium carbonate and some silica from seawater to construct their spicules. The bath sponge has a flexible skeleton of spongin and no spicules. The Calcarea or calcareous sponges comprise 500 marine species bearing spicules of calcium carbonate. The other groupings of sponges have siliceous spicules made from opaline silica. The coralline sponges have hard, protective calcareous skeletons and silica spicules. The sponges which secrete calcium carbonate contribute to the formation of reefs which turn into limestone.

Coral
Corals are cnidarian metazoans. They are also called coelenterates. Coral polyps harbour symbiotic algae. It is the algae that concentrate calcium bicarbonate dissolved in seawater into a supersaturated solution which precipitates out as calcium carbonate. Atlas & Bartha (1981, page 379) give the following equation:

$$Ca(HCO_3)_2 \rightarrow CaCO_3 + H_2O + CO_2$$

The calcium carbonate becomes a limy skeleton for the soft, living coral polyp. Corals are important rock builders. Their colonial habit leads to the formation of reefs in tropical seas. Many tropical islands are formed partly of coral limestone. Fossil coral reefs are found in what are now the Arctic and Antarctic regions today. Fossil corals are found from the Pre-Cambrian onwards and their presence is an indication of shallow water since corals only live down to 30 metres due to their symbiotic algae needing light for photosynthesis.
Barrier reefs form parallel to the shore line of land masses in shallow water. The Great Barrier Reef which runs for 2300 km off the north-east coast of Australia is composed of 359 species of hard corals which manufacture limestone, accompanied by many soft corals. The northward boundary of the reef seems to be the point at which the water gets too hot for corals to survive (Ince 2007, page 192). The Pacific and Indian oceans are studded with coral islands and atolls. An atoll is a ring of coral reef surrounding a lagoon.

Shell-bearing brachiopods and mollusks
The earliest shell-bearing animals to be clearly identified in the Cambrian period are the brachiopods. There are two main types of the sedentary brachiopods or Lamp Shells; the inarticulate brachiopods have the two shell valves composed of chitin and calcium phosphate, while the articulate brachiopods have shells composed of calcium carbonate. The inarticulate brachiopods were common in the Cambrian period and formed abundant fossils in shallow waters.

Phosphate-rich sedimentary rocks appear in the Cambrian so it was assumed that Cambrian seawater had exceptionally high concentrations of phosphorus. I think that the phenomenon might be better explained by the forms of life thriving in Cambrian seas. Conway Morris (1998, page 162) writes that during the Cambrian unusual amounts of sedimentary phosphate were deposited in the

148

shallow shelf seas that surround continents. The world's phosphate mines (used for fertilizer) extract this phosphate from rocks of Cambrian age – this indicates the massive quantities of phosphorus deposited.

Some geologists believe that there could be a direct correlation between the Cambrian 'explosion' and this episode of "phosphogenesis". I suggest a connection between the flourishing inarticulate brachiopods and the phosphate-rich sedimentary rocks of the Cambrian.

In mollusks such as gastropods and bivalves the mantle secretes a protein and calcium carbonate shell. The sedentary bivalves have contributed to some extent to reef-building and so to rock formation. The shells of the more solitary gastropods are found scattered on beaches, however, the constant abrasion of waves wears these away and they dissolve into the water after a few years.

The principle reef-builders are corals and algae including coralline algae. Oysters which are sedentary bivalve mollusks contribute to reef-building, as do Bryozoa which are small colonial filter-feeders.

In conclusion to this section, it is widely recognized that the shells and support structures of many protists, metazoans and some marine invertebrates accumulate as deposits which become sedimentary rocks. Mader (2007, page 390) cites radiolarians, diatoms, dinoflagellates, green algae (Chara) and foraminiferans as sedimentary rock forming organisms. Conway Morris (1993, page 220) mentions that protistan biomineralization – possibly of silica is known from the Precambrian (Riphean period). He cites as a reference, Kaufman, A.J., Knoll A.H. and Awramik, S.M. (1992) *Geology* Vol. 20, pages 181-185.

The most extensive deposition of sediments that have become sedimentary rocks has been performed by protist, often single-cell organisms living in the oceans. The protists include sedentary and motile photosynthetic algae and heterotrophic protozoans such as Radiolaria and Foraminifera.

As a summary of the list that I have given, the following rock types are formed by the following organisms:

Limestone Calcium carbonate ($CaCO_3$)	Cyanobacteria Foraminifera Green algae Red algae – reefs Sponges Coral with symbiotic algae – reefs Mollusks – oysters Bryozoa (colonial filter-feeder)
Dolomite Magnesium carbonate $(Ca, Mg)_2(CO_3)_2$	Green algae
Chalk (a type of limestone formed extensively in the Cretaceous period)	Golden algae / diatom coccoliths
Rock phosphates Contain PO_4 tetrahedra (Cambrian period)	Brachiopods
Chert Silica SiO_2	Radiolaria Diatoms (golden-brown and yellow-green algae) Silicoflagellates Sponge spicules – flint nodules

In favorable conditions these planktonic organisms can multiply rapidly and their large numbers makes their impact on the environment very significant. At the same time that photosynthetic algae were producing free oxygen and transforming Earth's atmosphere into an aerobic atmosphere, they were also busy depositing the chemical constituents of many sedimentary rocks. These chemical constituents can only be deposited in an aerobic environment where free oxygen is available.

4. Limestone and chert

To what extent are rocks produced by organic means?

Biologists show some enthusiasm for the subject, while geologists prefer to hedge their bets. As geology is currently presented, limestone is said to be formed by both organic means and non-organic physical processes, and even stromatolites are said to be both biologically formed and abiotic. Any rock with the label igneous rock such as granite is accounted for by physical processes alone, with at best a scarce mention of a possible sedimentary rock contribution. It is certainly unpopular to claim an organic origin for minerals and rocks, yet life, and especially microorganisms are so busy and so pervasive on Earth that I do not see how any understanding can be arrived at while ignoring them.

Limestone

Limestone would appear to be the least disputed rock of organic origin, especially when it contains the fossils of the organism which formed it. However, according to the current view, limestone is also thought to be formed by the Urey reaction.

In the Urey reaction carbon dioxide is removed from the atmosphere by becoming dissolved in rain water. This acid rain water reacts with silicate rocks on the Earth's surface. The end products of these reactions are magnesium or calcium-rich limestones which contain carbon dioxide extracted from the atmosphere. If the limestones are subsequently buried deep within the Earth's crust and heated up, the process is reversed and carbon dioxide is released back into the atmosphere.

It is thought that as mountains such as the Himalayas are pushed up and erosion and weathering take place, the Urey reaction transfers atmospheric carbon dioxide to the weathering products of rock which are washed into the oceans as sediments. It has been found, however, that if the Urey reactions are as effective as claimed, the atmosphere would have been stripped of carbon dioxide long ago (Lamb & Sington 1998, page 163-166).

Thus, according to the theory of chemical weathering named the Urey reaction, carbon dioxide from the atmosphere enters into a reaction with silicate rocks to produce carbonate minerals. The alternative view is that living organisms extract carbon dioxide from the atmosphere as part of the chemistry of life. Biological processes employ carbon dioxide to produce calcite that is used to make structures which serve a protective function for the organism.

Most of Earth's carbon dioxide is locked up in sedimentary rocks of the crust as huge accumulations of limestone. This has changed Earth's climate which has mostly been much milder than at present. The sequestering of carbon dioxide mediated by living things (according to my belief) has produced a cooler planet. The mass of carbon contained in living organisms on Earth today is over four times the atmospheric mass of carbon (as carbon dioxide). The mass of carbon in rocks – there due to the activity of living organisms is 100 000 times this (Lamb & Sington 1998, page 233).

A much larger proportion of Earth's crust, however, is formed by silicate minerals than by carbonate minerals. Basalt composed of feldspar, pyroxene and olivine consists of aluminosilicates which originate in the Earth's mantle. Granite is largely composed of silicates and is a rock of very mixed and varied crystalline components. Mica is formed of sheet silicates that have a flaky texture and are a major constituent of schist. Quartz is a crystalline silicate that appears in many rock formations and is a component of many rock types. It is to this last silicate mineral that we will now turn our attention.

Quartz or chert

The terms quartz, chert, flint and chalcedony refer essentially to the same thing, and the labels are often used interchangeably. The labels have histories associated with different countries and industries. Quartz or chert is the most abundant mineral in Earth's continental crust. It is formed of a lattice of silica (SiO_2) tetrahedra.

Transparent quartz is macrocrystalline and has a crystal shape of a six-sided prism. Quartz crystals are known over a metre long. Opaque quartz is microcrystalline or cryptocrystalline. Chalcedony is an opaque type of quartz. Sand is quartz grains coated in hematite. Silica glass (SiO_2) is formed by lightning strikes in quartz sand.

Chert varies in colour from white to black, but is most often grey or brown. When it contains oxidized iron it is red, and when it contains reduced iron it is green. Chert occurs as oval to irregular nodules in greensand, limestone, chalk and dolostone formations. Where it occurs in chalk, it is usually called flint. Chert or flint nodules often contain fossils. Chert is also laid down as thin beds, as in the case of jasper and radiolarites. Thick beds of chert occur in deep geosynclinal deposits.

Quartz gemstones (SiO_2) include amethyst, citrine and cairngorm. Quartz crystals coloured by iron oxides include jasper, cornelian, onyx and agate.

Quartz or chert is highly resistant. It resists weathering, recrystallization and metamorphism. We see eroded examples of quartz or chert all around us in the form of cobble stones, pebbles, gravel and sand. These stones have been formed by the action of waves along beaches, stream torrents and river currents.

Sheringham beach
Sheringham is a small town on the coast of East Anglia, England. When I went down to the beach adjacent to the town in 2009, I observed that to the east the soft chalk cliffs were being eroded by the sea leaving on the beach a white slippery shelf embedded with large and small flint nodules. The fresh flint nodules were being released from the chalk and lay strewn about the beach, where they were being broken to pieces by wave action. The angular broken pieces revealed the fossils of white calcareous sponges inside the nodules as well as microscopic fossils detectable as a change in colour inside the flint.

The angular pieces were being eroded by wave action into rounded cobble stones. Further along the beach to the west, the cobble stones had become pebbles, further along still the pebbles had become gravel, and finally the small sized gravel had become sand along the shore line. The sand is retained by groins dividing the beach into sections.

It appeared that the sand, gravel, beach pebbles and cobble stones had come from erosion of the flint nodules from the cliffs. I was astounded to surmise that the flint nodules had almost certainly been formed by sponges living in the sea. Thus, the clear gradation of material lying on the beach originated ultimately from the activities of a metazoan animal!

I then started thinking about what this meant. My observations of the fossil sponges inside flint nodules were confirmed the following week when a question on QI (a BBC quiz show on science) gave the answer that the chert or quartz of flint stones is secreted by the sponge itself found inside. (Some sponges have spicules composed of silica and a body wall composed of calcium carbonate).

If flint nodules composed of chert could be produced by a lower metazoan animal, I wondered whether beds of chert and seams of chert could have been produced by some sort of unicellular organism since these were extremely abundant over the eons of Earth history.

Microfossils found in chert
It is well-known that chert is often associated with fossils, many of which are microscopic. It is generally explained that chert due to its fine texture is ideally suited to the preservation of fossils. In other words, the mineral was precipitated around organisms that by chance were present in the environment. The chert is thought to have been formed by the physical process of chemical precipitation. Lamb & Sington (1998, page 177) writing about the Barberton Greenstone Belt of South Africa and Swaziland state, "The cherts formed when silica was precipitated from hot springs in a region of intense volcanic activity."

I formed the view, however, that the organisms found in chert, like the organisms found in limestone and the sponges found in flint nodules, could be intimately associated with the production of the mineral itself.

This is a list of the microfossils found in chert:

- The Fig Tree Formation chert in the Barberton Mountains between Swaziland and South Africa is 3200 million years old. It preserves non-colonial unicellular bacteria-like fossils.

- The Apex Chert of the Pilbara craton, Australia is 3400 million years old. It preserves eleven taxa of bacteria.

- The Gunflint Chert of western Ontario is 1900 to 2300 million years old. It contains fossils of Archaea, cyanobacteria, organisms resembling green algae and fungus-like organisms.

- The Bitter Springs Formation of the Amadeus Basin, Central Australia is 850 million years old. It preserves cyanobacteria and algae.

The chert beds of the Bitter Springs Formation have been found to contain 30 species of microfossil, including diverse cyanobacteria, chlorophycean algae, and possible dinoflagellates, fungi and heterotrophic bacteria. The larger cells contain internal structures and have been assigned to green algae. The formation is composed largely of dark limestone and dolomite, and contains black chalcedonic chert in fine laminated layers. The formation contains numerous stromatolites (Schopf 1968).

Looking at this example, it is conceivable that the limestone and dolomite of the Bitter Springs Formation was formed by green algae, the layers of black chert by dinoflagellates and the stromatolites by cyanobacteria. Consequently, different microscopic life forms would have contributed to the Formation with different minerals. I accept that some of the organisms present such as fungi simply happened to be members of the biocommunity and were not instrumental in mineral formation.

In the description of the Gunflint microbiota by Preston Cloud (Cloud 1988, pages 239-241) he observes that the microfossils are found in finger-like stromatolites, not in the black chert between branches. The formation is described as stromatolitic chert. The Gunflint filaments resemble cyanobacteria (which Cloud calls 'proalgae') with terminal resting cells, nitrogen-fixation cells and oxygen-shielding cells. He notes that these bacterial filaments have been preserved by being sealed inside the silica chert. The Gunflint Chert Formation is associated with the Gunflint Iron Formation. The Gunflint microfossils were first reported in 1954, but were not accepted as fossils until Preston Cloud published on the subject in 1965.

My overall conclusion is that the cyanobacteria preserved in these formations has been responsible for the production of the limestone stromatolite component of the formation. Green algae may also have produced limestone, while flagellated unicellular algae (dinoflagellates or diatoms) would have been responsible for producing the chert component. Since this planktonic algae with its silica shells would have rained down from surface waters to the bottom, it would have encased the organisms and formations already found on the seabed. It would also have encased bacteria living on iron and other energy sources on the seabed.

It is my opinion that large quantities of chert or quartz would have been laid down in layers at the bottom of oceans from flagellated algae carried by the tides and dying off seasonally. This would have occurred throughout Earth history from about 2.5 thousand million years ago. These layers are sometimes found intact as components of continental crust, but more often the broken up remains of them are the only evidence. Quartz layers when disturbed shatter into small particles which form sand. Sand is the principle component of redeposited sedimentary rocks such as sandstone.

Banded Iron Formations (BIFs)

Banded Iron Formations (BIFs) belong to the Precambrian age. They are associated with some of the oldest rocks on Earth and are commonly found in ancient greenstone belts.

BIFs are composed of alternating layers of iron oxides and chert or quartz. The iron often appears as concentric rings. The layers of iron oxide may be millimetre to centimetre thick. BIFs can be hundreds of metres thick and hundreds of kilometres wide. At Pilbara in northwestern Australia the chert layers with banded iron formations are 1000 metres thick (Cloud 1988, page 155-157).

According to current thinking the chert layers of BIFs came about by the precipitation of silica from sea water. The BIFs with concentric rings resemble dried versions of hot, bubbling mud pools found in areas of volcanic activity and so have been associated with volcanic activity.

The iron oxide layers in BIFs have been associated with the advent of molecular oxygen in the waters and atmosphere of early Earth giving rise to the spontaneous oxidation of reduced ferrous iron that was dissolved in the waters. Preston Cloud (1988, pages 235, 241-247) formulated the hypothesis that this free oxygen was released by the photosynthesis of cyanobacteria that colonized the Earth at this early date.

It has been pointed out that banded iron formations could only have formed if the Earth's atmosphere (including its waters and oceans) contained virtually no free oxygen. The layered structure of BIFs shows that oxygen was only available seasonally for the oxidation of iron, interspersed by periods of no oxygen being available. If there had been an aerobic atmosphere containing a constant amount of oxygen, the iron oxides would not appear in layers.

Banded iron formations mostly date from between 2.5 and 1.8 thousand million years ago. Nothing like it is found in younger rocks. The reason for this is that when aerobic conditions began to prevail, the oceans no longer contained ferrous iron in quantity and so BIFs ceased to be formed.

Reflecting upon the existence of banded iron formations as an indication of what early Earth was like I realized that if photosynthetic unicellular algae such as diatoms and other silicoflagellates had been living in the shallow seas or pools of early Earth, then there would be one simple explanation for banded iron formations. The photosynthesizing blooms of planktonic algae would release oxygen into the water each summer. In generally anaerobic conditions this free oxygen would cause ferrous iron dissolved in the water to be oxidized. The precipitated iron oxides would sink to form a layer of iron oxides on the seabed. In the cold of winter the algae would die off and sink to the seabed where the siliceous tests (tiny shells of diatoms) would form a layer composed mainly of silica. Over time the silica would be compacted into chert layers. This seasonal cycle would be responsible for the banded nature of these iron formations.

The first occurrence of banded iron formations at 2.5 thousand million years coincides with dating of the oldest known acritarchs. Acritarchs represent the resting stage of foraminifera, green algae and flagellated algae. The discovery of some fossils in an iron formation also provides a link between algae and banded iron formations from an early date. Centimetre long tube-shaped fossils of *Grypania* represent some of the oldest fossils of eukaryotic algae. They have been found in the

2.1 thousand million year old iron formation in Michigan, North America (Han & Runnegar 1992).

I have reached the conclusion that the oxygen-producer had to be silica-producing algae and not limestone-producing cyanobacteria to account for the chert layers in BIFs.

Therefore, whereas before I accepted that the Earth's aerobic atmosphere was produced by cyanobacteria; I would now add that it was also produced by photosynthetic algae. Also, whereas before I accepted that the earliest date at which eukaryotic single-cell organisms could have existed on Earth was 1.8 thousand million years ago, I now revise that date to 2.5 thousand million years ago.

5. Granite

All the terrestrial planets and satellites of the solar system have surfaces of basalt; granite appears to be unique to the Earth. Granite has always been classified as an igneous rock formed from molten magma, thus having a purely volcanic origin. Granite may, in fact, contain a large proportion of melted sedimentary rocks – this would explain its composition and its occurrence only on Earth.

Granite intrusions
It appears that where molten magma seeks a way to the surface beneath the sedimentary strata of continents, it partially melts the sediments it encounters and becomes mixed with them as it pushes its way up. The result is a rock of visibly varied composition. Granite is composed of crystals of quartz, feldspar and mica.

The magma forming granite does not reach the surface and erupt like lava from volcanoes, but pools up beneath layers of sediments forming a subsurface mass known as a granite intrusion. Mountain chains on land are formed to a large extent out of granite intrusions. The granite is only exposed at the surface by the weathering away of overlying rock.

It is now recognized that granite is produced by the melting of crustal material at depth (Ince 2007, page 86).

The main difference between basalt and granite is that granite is enriched in silica (SiO_2) or quartz. Basalt has less than 20 % quartz, while granite has at least 20 % quartz. Quartz is pure silicon and oxygen. The other silicate minerals in granite are composed of silicon, oxygen and metals; these silicates include feldspars, olivine, amphiboles and pyroxenes. Mica consists of flat-shaped crystals.

Andesite

The lava erupted from volcanoes on land, as well as from volcanic arcs in the ocean is generally different from the lava erupted from mid-ocean ridges under the oceans. The lava erupted on land is a sticky lava enriched in quartz called andesite. Andesite gives rise to granite.

Andesite lava erupting from subduction volcanoes is likely to be produced by the melting of marine sediments drawn into the mantle along subduction zones. The mixing of basalt magma and 'siliceous oozes' formed from the tests of diatoms and silicoflagellate algae accumulated on the seabed would explain the composition of this type of lava.

Therefore, there are two sources of granite; one is from andesite lava erupted from subduction volcanoes and the other is from magma rising beneath landmasses by forming dykes along fractures formed by fault systems in areas of volcanic activity. The solidification of this magma in underground channels forms granite intrusions in mountainous regions.

When magma is intruded into sedimentary rocks and melts its way up as it pushes towards the surface, it may encounter limestone rocks or layers of chert or quartz. If it encounters limestone, the heat of the magma may cause calcium carbonate to dissociate into calcium oxide and carbon dioxide, or to recrystallize into calcite. If it encounters quartz layers, the quartz will melt and mix with the basalt magma and later recrystallize. This enriches the magma in quartz and makes it increasingly sticky.

Therefore, granite can be classified as a rock of dual origin – igneous and sedimentary. It could be classified as a metamorphic rock. The sedimentary component of granite makes it a rock that is partly of organic origin.

6. Metal ores

Gold
Gold is a noble metal which means that it can occur in its elemental
form as pure gold. It does not react readily with other elements; it
does not tarnish or corrode. However, gold is also a transition metal
since it can form a few compounds. The gold compounds of practical
importance are gold(I) chloride AuCl; gold (III) chloride or gold
trichloride $AuCl_3$; and chlorauric acid $HAuCl_4$. These compounds are
involved in the electrolytic refining of gold.

Pure gold is very soft so it is usually alloyed with other metals to
increase its hardness for use in jewellery, gold ware and coinage. It is
alloyed with silver, copper, zinc, nickel and platinum. The colour of
these gold alloys goes from yellow to white as the proportion of silver
in them increases. The gold content of the alloy is expressed in 24ths
called karats. A 12 karat gold ring has 50 % gold, while a 24 karat ring
is pure gold.

Gold occurs in rocks as invisible disseminated grains and more rarely
as flakes large enough to be seen. It is even rarer for gold to be found
as masses or veinlets.

'Panning for gold' is the extraction of nuggets of gold from alluvial
deposits (river beds and dried up water channels). The gold of alluvial
deposits comes from grains of gold being eroded from rocks by rivers
and welded together into nuggets by water action.

When gold is found in situ (rather than as an alluvial deposit), the
invisible grains of gold or tiny flakes are associated with quartz. The
quartz ore usually also contains copper, lead and iron pyrites (FeS_2) or
'fool's gold'.

The origin of quartz ores bearing precious metals is not fully known.
It is thought that hydrothermal super-heated waters associated with
volcanic activity carried up the gold in partial solid solution from the
Earth's mantle with other minerals. The hydrothermal water
penetrated fissures in the rocks of Earth's crust and as it cooled the
quartz precipitated out and the cargo of metals were deposited in the
fissure as a physical process. As the quartz was precipitated from the
water, quartz crystals were formed; some quartz crystals have been

found several metres in length. To reflect this thinking, these deposits are called 'hydrothermal veins'.

It has also been noted that gold is often associated with not only quartz, but also carbon of biological origin. This suggests the participation of some form of life in the accumulation of these metals as ores. Ancient greenstone areas such as the Barberton Greenstone Belt of Transvaal South Africa and Swaziland, and the Pilbara region of Western Australia have been gold rush areas. These areas also have copper ore deposits. Bacteria-like filaments have been found in the Pilbara gold fields, 3.5 thousand million years old. The presence of fossil bacteria in some gold mines has led Margulis & Schwartz (1982) to suggest a microbial participation in the concentration of certain metals:

"Gold in South African mines is found with rocks rich in organic carbon, associated with fossil bacteria and probably of microbial origin. In Witwatersrand, the miners find the gold, deposited apparently more than 2500 million years ago, by following the "carbon leader" that leads people to the gold. Copper, zinc, lead, iron, silver, manganese, and sulfur all seem to have been concentrated into ore deposits by biogeographical processes that include bacterial growth and metabolism." (Margulis & Schwartz 1982, page 42).

Heavy metal chelation by bacteria
Some bacteria take in heavy metals that are toxic to other forms of life. Heavy metals such as mercury, lead and arsenic are taken into the cell by methylation which means that carbon-metal bonds (rather than sulphur-metal bonds) are formed making an organometal. The process is called chelation.

The relatively high concentration of heavy metals in sewage sludge is due to the ability of microorganisms to concentrate these pollutants. Bioremediation of water is the use of bacteria to clean pollutants from water. Bacteria will even absorb heavy metals such as uranium and plutonium. When heavy metals are concentrated in the bacterial cells, if the biomass is reduced by production of methane, the heavy metals are released and concentrated. This is useful for removing these dangerous radioactive elements from water.

Bacteria must derive some benefit from concentrating heavy metals within their cells, but it is not known what the benefit is. It is known that bacteria fall prey in large numbers to viruses called bacteriophages. The heavy metal may protect the bacterium from viral attack.

Algal concentration of metals

It seems to me that the association between quartz, organic carbon and gold specks points to the accumulation of gold as well as of copper and lead by a species of algae with a silica test.

I propose that the quartz of quartz ores originated as a deposit formed from the tests of silicoflagellate algae and that the gold specks represent gold incorporated into the algal cells – a certain species of algae forming a concentration of certain metals. The algae may have been practicing the chelation of metals to protect itself from water-borne mildew or bacteria.

At a later stage, the deposit may have undergone metamorphic changes through heating from volcanic activity causing the quartz to melt and recrystallize with the gold still embedded in it. Or the circulation of hydrothermally heated water may have dissolved the quartz deposit and redeposited it in veins in fissures in the rock as secondary quartz. The crystallized secondary quartz retained its metal content. (Secondary quartz will be discussed in the next section under Quartz: chalcedony).

The carbon found associated with quartz ores allegedly represents bacteria or the remains of bacteria. It is possible that these bacteria grew upon the organic remains of the algae.

In support of the organic formation of metal ores are the following observations:

In the processing of ores, microorganisms are sometimes used as biocatalysts in the recovery of precious metals from dilute-process streams. Green algae are used to concentrate precious metals including gold and silver within their cells, and thus separate the precious metal from the other material in the deposit.

Gold is apparently not toxic and metallic gold can be added to food for human consumption with no harmful effects. Copper is used by humans as a germicidal agent – cuprous oxide is used as a fungicide;

162

cuprous chloride as a disinfectant; and cuprous sulphate as a fungicide and bactericide.

Copper is present in the ashes of seaweeds and in many sea corals. (Seaweeds are algae and corals contain symbiotic algae).

Silver is often present in ores containing copper, lead and zinc. Silver also occurs naturally in many minerals. Silver is toxic to bacteria, fungi and viruses. At one time silver was incorporated into clothing to prevent bacterial and fungal infection. It is possible that silver was concentrated by algae to prevent bacterial and fungal attack.

These observations show that algae do take gold and silver into their cells. The reason for this is not known, however, it is known that silver and copper kill bacteria and fungi or mildew and so there is the possibility that algae incorporate these metals as protection against attack by these microorganisms.

7. Rock types

The following list is a simplified classification of the rock types which make up the Earth's crust with a discussion on the origin of each type.

Rocks formed from molten magma

Basalt
Basalt emerges from the Earth's mantle as molten magma and solidifies into volcanic lava. Basalt forms the ocean floor and on land it erupts from volcanoes forming volcanic cones and occasionally outflows called flood basalts.

Basalt is composed of various silicate minerals including aluminosilicates. These silicates consist of silicon, oxygen and a variety of other elements such as magnesium, iron, calcium, aluminium, sodium, potassium, barium and titanium. The forms that basalt takes and its composition has been described in Chapter 18, section 7. There are different types of basalt according to composition.

Granite

Granite is an abundant component of continental crust. It is a coarse-grained crystalline rock of which the colour varies from pink to grey to black. Granite is formed from the slow cooling of magma below the surface of the Earth. It is only exposed later by erosion in mountainous regions.

Granite contains at least 20 % quartz which is pure silica (SiO_2). It also contains the silicate minerals found in basalt including feldspar which is an aluminosilicate. Granite viewed under the microscope shows a mosaic of large-sized crystals that may be a mixture of quartz, feldspar and mica.

In section 5 of this chapter I discussed the general new consensus that granite is formed by magma pushing up beneath landmasses through layers of sedimentary rocks and partially melting these sedimentary rocks which become mixed with the basaltic magma.

The magma forming granite does not reach the surface since it is capped by the overlying sedimentary layers. It accumulates below surface as a basement rock with intrusions into overlying sedimentary rocks. Granite forms a major part of continental landmasses. Granite has been intruded into the crust during all geologic periods, although much of it is of Precambrian age.

Bacterial transformation of primary minerals

Mica and clay

Micas are sheet silicates commonly containing aluminium, potassium, sodium, magnesium and iron. Water is trapped between the sheets of silicates. Large chunks of mica are called books because the sheets resemble many pages. The flexible sheets or lamellae tend to flake off easily. Because the sheets of mica can be transparent and are highly heat-resistant, mica sheets have sometimes been used instead of glass for small windows, for example, as peep-holes in boilers and lanterns.

Mica is produced by the weathering of basalts; the continued weathering of mica produces clay which is a layered silicate.

In Chapter 18, section 7 I discussed the new research into the microbial weathering of basalts on the deep ocean floor. There is convincing evidence that secondary minerals including mica and clay which fill fractures and spaces in the ocean crust as well as iron-oxide crusts and ferromanganese nodules found on the seabed are produced by chemolithotrophic bacteria obtaining chemical energy from the reduced compounds contained in basalt.

Diverse communities of bacteria which appear to have an interdependent mode of existence have been cultured from basalt formations in the vicinity of mid-ocean ridge systems. The volcanic activity below these ridge systems would supply the heat necessary for bacterial metabolisms to function at a higher rate, although much of the basalt weathering occurs at low temperature. The vesiculated form of basalt rocks and the high permeability of the uppermost 200 to 500 metres of basaltic ocean crust appear to facilitate the entry of microorganisms which live as endolithic colonies.

The potential for the microbial weathering of the basalt ocean floor into mica and clay is very high. However, due to the difficulties of conducting research on the deep ocean floor, the scale of impact of these processes of biomineralization is not known.

Basalt and granite found on land are subject to chemical weathering which changes them into clay and chemicals in solution. Quartz is not weathered in this way since it is highly resistant.

It is now known that chemical weathering often involves the production of acids by microorganisms. Fungi found on rock surfaces have been shown to decompose silicates by producing citric and oxalic acids.

Vermiform transformation of secondary minerals

Mudstone and shale
Clay is a valuable source of elements vital for life. Bacterial lithotrophs appear to be the key to the initial decomposition of basaltic rocks and volcanic glass into mica and fine clay. The fine clay produced is an ideal substrate in which lower invertebrates such as marine worms can live. Many types of worms ingest sediments of clay mixed with organic matter digesting the organic matter and probably obtaining vital elements from the clay itself. Worms expel the remaining silicate sediments and organic matter from their guts as

mud. Mud on the seabed is produced by marine worms, the way soil is produced by earthworms on land.

On land silicate rocks and volcanic ash from volcanic eruptions undergo chemical weathering by acids produced by fungi and bacteria leading to the formation of clay. These clays are a substrate for plants providing them with necessary elements. Since plants are consumed by animals, mineral elements are passed into the ecological food web by this means.

It is likely that marine and terrestrial worms which ingest sediments as deposit-feeders also absorb elements contained in clay minerals, and in this way also pass mineral elements into the ecological web of life.

The muddy seabed is recurrently buried by the accumulation of sediments on top of it, and layers of mud accumulate until over the course of geological time they turn into shale through compression. During the process of lithification, the pressure to which the mud is subjected causes the grains to form layers. Mudstones and shale have visible layers, whereas mud does not have any structure.

In the classification of sedimentary rocks, a rock composed of fine clay-sized particles (less than 0.0039 mm) is called a claystone. Silt has a slightly larger particle size (between 0.0039 and 0.063 mm) and forms siltstone. Mixtures of clay and silt form mudstone. A laminated mudstone is shale.

The sedimentary rocks on land include extensive areas of shale. These rocks would once have been the muddy bottom of shallow seas inhabited by large communities of marine worms. The muddy material would have been mixed in and expulsed from the guts of these worms. Later on the mud transformed into shale became uplifted above sea level by folding and faulting to become part of the sedimentary rock strata forming landmasses.

Black shale from the Palaeozoic and Mesozoic eras are common. These shales are black because they contain unoxidized carbon and were formed in anoxic stagnant waters. Shale often contains the imprints of animal tracks and burrows as well as fossils. The tracks and burrows were those of marine worms and marine arthropods living on the seabed.

If vermiform transformation of secondary minerals is instrumental in the production of shale, then no shale should be older than the Cambrian period starting at 570 million years ago when multicellular bilateral animals first appeared (or at most 600 million years ago just prior to the Cambrian).

Metamorphic rocks known as schist and gneiss may date from earlier than this. Schist is formed from mica and other flaky minerals while gneiss is banded but lacks flaky minerals. Schist and gneiss are likely to contain metamorphosed clay rather than shale as is often stated. The exact original composition of gneiss is unknown since it has been completely transformed by recrystallization. (Schist and gneiss are described under the section on metamorphic rocks).

I have found a number of vermiform candidates for making mud in *Grzimek's Animal Life Encyclopedia* Volumes 1 and 2. First I will list the worms and assorted odd organisms which live in marine sediments:

- Rotifera: inhabit sediment and feed on suspended organic particles, microalgae, ciliates (protists) and bacteria.

- Gastrotricha: live between sand grains and feed on diatoms, foraminifera, protists and bacteria.

- Loricifera (Girdle wearers): live in marine sand and mud and feed on organic particles, microalgae and bacteria.

- Gnathostomulida: live in detritus-rich marine sand and graze on fungal threads and bacteria.

- Micrognathozoa (Jaw animals): inhabit the sandy bottom and feed on diatoms, cyanobacteria and bacteria.

- Adenophorea (Round worms): live in marine sediments and feed on fungi and bacteria.

- Turbellaria (Flat worms): many species live between grains of sand. There are predators and scavengers feeding on protozoans, copepods, small worms and mollusks.

This second list includes large sized worms which ingest large quantities of sediments, often leaving worm casts on the seabed:

- Sipuncula (Peanut worms): most are deposit-feeders which ingest sediments. They live in semi-permanent burrows in coarse and silty sand or in crevices under rocks. Peanut worms have an introvert feeding apparatus with hooks and tentacles. They ingest sediment and associated biomass by collecting it with their tentacles.

- Polychaeta (Clam, Sand and Tube worms): there are planktonic and benthic (sea bottom dwelling) polychaetes which are annelid worms. The benthic polychaetes live in the intertidal zone close to the shore as well as in the deepest depths of the ocean in sandy and muddy sediments. They have all types of feeding habits and there are free-moving forms and sedentary tube-dwelling forms. Their size varies from a few millimetres to three metres long. The non-selective deposit-feeders ingest sand or mud grains showing little discrimination for the size and nutritional value of the particles, assimilating any organic matter in the ingested sediment. Selective deposit-feeders utilize palps, tentacles or buccal organs to select particles with high nutritional value. Deposit-feeders have a sac-like pharynx that sucks in the sand and mud grains from the sediment when burrowing. Polychaetes play an important role in turning over bottom sediments in the sea.

- Hemichordata (Acorn worms): these solitary worms live in U-shaped burrows remaining there for shelter. They collect sediments with a proboscis from one end of the burrow and evacuate it at the other opening to the burrow forming large worm castings several inches high a short distance away from the burrow. Some Acorn worms are 2.5 metres long.

Minerals formed from the tests or shells of living organisms

Many organisms form shells or tests for protection by secreting mineral building materials. These, often tiny organisms, have lived and do live in vast numbers in the Earth's oceans and fresh waters. As these organisms flourish, multiply and die, their shells accumulate on the bottom of oceans and lakes, and the layers eventually become compacted into rock. Therefore, the rock formation consists of the fossil remains of life itself.

Limestone
Limestone consists of the organically-made mineral calcium carbonate ($CaCO_3$) derived from the tests of protists such as foraminifera, limestone-forming algae and the protective structures of animals such as corals.

Limestone consisting of corals was formed in warm, shallow seas – coral reefs only grow in shallow water. The limestone may contain the remains of the reef in situ, or consist of the ground-up pieces of reefs which were broken up. These limestones have visible fossils embedded. Finer limestones are formed from marine algae.

Chalk which is also composed of calcium carbonate is formed from the coccoliths of planktonic golden algae. The coccolith was the resting stage for these motile algae.

Quartz
Quartz composed of silica (SiO_2) is the second most abundant mineral in the Earth's crust after feldspar. It occurs as a component of granite and it is present in many sedimentary rocks. **Chert** is a microcrystalline form of quartz.

Quartz is both highly resistant to mechanical weathering and inert to chemical weathering and so it is not transformed into other secondary minerals. Quartz, however, is easily shattered and so it tends to accumulate as sandstones and other detrital rocks.

The solubility of quartz in water at 25 °C is only about six parts per million. However, river water in temperate climate regions contains about 14 parts per million of silica (*Encyclopedia Britannica 2011 Standard Edition*: Silica mineral). This transport in solution means that secondary quartz can serve as cement in many sedimentary rocks, encasing the grains and holding them together.

Chalcedony is a white or whitish, finely crystallized quartz that forms rounded crusts and rinds in volcanic and sedimentary rocks. Chalcedony is formed by gases and solutions saturated with silica and other compounds escaping from molten andesite (the magma of granite) through deep fractures. As the gases and hydrothermal water cools, quartz is precipitated along the channel ways to form veins. Chalcedony may be deposited in this way, or large crystals of colourless quartz may form called rock crystal.

Stalactites in caves are made from chalcedony.

Flint is a form of chert found as nodules within chalk deposits. Flint consists of silica secreted by sponges which once lived on the seabed in shallow waters. Sponges remain within the flint nodules as fossils. Flint is classified as an organic sedimentary rock especially since it is associated with easily recognizable fossils.

There is another form of quartz called **bedded chert**. Bedded cherts occur as individual bands or layers ranging from one centimetre to several tens of metres in thickness. They are associated with submarine volcanic flows and with deep-water mudrocks. Many bedded cherts are composed almost entirely of the silica tests of diatoms and radiolarians. Chert is produced by compacting and recrystallizing the unconsolidated siliceous ooze deposits that continue to accumulate on the floors of oceans. The modern siliceous oozes accumulate in latitudes where there is high organic productivity in planktonic algae and protozoans in warm surface waters. When the organisms die their tiny tests drift down and settle on the abyssal floor of oceans far from continental landmasses. In these deeper parts of the oceans calcareous oozes do not occur. Over time this gives rise to mudstones with beds of chert. It is thought that some bedded cherts which are associated with pillow lavas from submarine volcanic activity, are derived from precipitated silica gels, rather than of organic origin (*Encyclopedia Britannica 2011 Standard Edition*: Sedimentary rock: origin of cherts).

It is my contention that the contribution of organically produced bedded chert has been far greater than recognized up to now. I believe that bedded cherts have been produced by protist planktonic life living in seas and oceans from 2.5 thousand million years ago onwards. These Precambrian beds of chert formed at the bottom of seas before continental landmasses existed would be the main source of quartz now found incorporated in other rocks. Basaltic magma pushing up through bedded chert would become granite, one of the foundation rocks of landmasses. Volcanic activity also involved the melting of quartz and the redeposition of it in solution by hydrothermal fluids.

Extensive faulting has shattered other bedded chert layers. Disrupted chert beds have given rise to sand which has become consolidated into sandstone and other detrital rocks.

Therefore, in my classification quartz comes under the heading of 'Minerals formed from the tests or shells of living organisms' although there has been an extensive redeposition of quartz by physical processes.

Redeposited sedimentary rocks
Redeposited sedimentary rocks are formed from the weathered or eroded particles of a former rock or mineral deposit. The main agents of erosion are water and ice. The rocks of landmasses are constantly eroded by rain and ice splitting rocks apart; rivers and glaciers abrading rocks and grinding them away; and waves along coastlines cutting wave platforms with cliffs constantly receding away from the sea.

The eroded rock fragments are transported away by rivers and sea currents, and occasional floods and redeposited as sediments. The sediments are converted to solid rock by a cementing process called lithification. Water percolating through the sediments may cement the deposit together by the deposition of calcite or silica carried in solution or by the conversion of soluble ferrous iron to insoluble ferric iron with the precipitation of iron oxides.

Rock formed from the broken-up pieces of a former rock is known as clastic rock. **Shale** is a clastic rock formed from fine clays and silts. **Sandstone** is composed of sand-sized grains of quartz, while breccia has larger, sharp particles and conglomerate has larger, rounded particles.

Sandstone is cemented together by iron oxide (hematite). The grains of sand made from quartz are transparent, but sand stone is yellow, brown or red depending on the quantity of iron oxide around the sand grains. Sandstone is very common.

Red Beds are sedimentary rock formations formed from quartzites. A **quartzite** is sandstone that has been converted to solid quartz rock by the precipitation of silica between the grains or by metamorphism welding the sand grains together. The oldest extensive Red Beds are 1.8-2 thousand million years old. Red Beds are found in basins.

The Precambrian Red Beds show that quartz was abundant at this early date and that the level of free oxygen in the atmosphere and waters had caused the Earth to rust. I would associate these Red Beds with pools of water in basins or craters on the early Earth inhabited by photosynthetic silicoflagellate algae. The algae would form the deposit of quartz and later faulting and folding would break the quartz layers into sand. Later Red Beds may have formed in inland basins associated with deserts. It has been claimed that Red Beds indicate the former existence of deserts and hot climatic conditions

Some sedimentary deposits have been wind-blown rather than water deposited. **Loess** is composed of silty and loamy material loosely consolidated; it is wind-deposited and covers 10 % of the land surface of the Earth.

Volcanic ash from erupting volcanoes is composed of basalt rock and glass. The fine volcanic particles travel as high ash clouds over large areas until fallout occurs. Pyroclastic ash beds may be converted into **bentonite** sedimentary rock by the alteration of volcanic glass into clay minerals. The decomposition of nutrient-rich volcanic ash is responsible for some of the world's best soils.

Heat and pressure transformed rocks

Heat from volcanic activity may cause localized or more widespread metamorphism. Sedimentary or igneous rocks may undergo recrystallization when subjected to temperatures which may be between 300 and 1000 °C. If the rock is also subjected to pressure, the crystals may realign themselves according to the orientation of the stress when it is coming from a specific direction.

Metamorphic rocks are harder than the original sedimentary rocks from which they were formed. Metamorphism preserves ancient sedimentary rocks and basalt by making the minerals in them more resistant to weathering. An example of this is the basalts and sedimentary rocks of Greenstone belts which date from the Precambrian.

One of the best-known and most appreciated of metamorphic rocks is **marble**. Marble results from the metamorphism of limestone ($CaCO_3$) or dolomite ($(Ca, Mg)_2(CO_3)_2$). The swirls and veins in marble are produced by impurities of clay, silt, sand, chert and iron oxides.

Another common metamorphic rock is **slate**. Metamorphism converts shale into pelites. If the layered structure of the original shale is retained and it is possible to split the rock along its bedding planes, it is called slate.

Gneiss is a term used for metamorphic rocks which have a banded, folded structure and are coarse-grained having undergone recrystallization. Some types of gneiss (pronounced 'nais') are thought to come from metamorphosed granite while other types of gneiss are thought to be descended from sedimentary rocks.

Schist is a foliated rock containing flaky scales. Schist is derived from the metamorphism of mica and other lamellar minerals such as chlorite, talc, hornblende and graphite. This flaky mineral is probably derived from basalt that has been transformed into mica and clay. Another component of schist is sandstone.

Precipitated sedimentary rocks

Evaporate minerals are salts precipitated out when a body of water dries out. When seas dry out **rock salt** or halite which is sodium chloride (NaCl) is precipitated out as crystals. The salt we eat comes from ancient seas or from the present sea. When freshwater lakes dry out, **gypsum** which is calcium sulphate ($CaSO_4.2H_2O$), may be left as a deposit of clear crystals. Barite which is barium sulphate ($BaSO_4$) and nitrates are also precipitated out of lakes.

Precipitated rocks are known as evaporites or chemical sedimentary rocks.

Buried organic matter

Coal is considered to be a type of sedimentary rock. As noted in Chapter 19 section 7, Precambrian coal is formed from the buried remains of cyanobacteria, photosynthetic algae and fungi; Carboniferous and Permian coal is formed from the first forests of primitive trees that covered low lying landmasses; and Cretaceous coal is formed from peat accumulated from forests of seed-bearing trees.

Coal-bearing rocks are called **coal measures**. The seams of coal may be several metres thick with layers of coarse and fine sandstones, muddy rocks and sometimes limestones in between. Most coal was laid down as peat in swamps which were buried by sediments brought by flood waters.

Crude oil reservoirs are sedimentary basins of porous sedimentary rock saturated in crude oil or petroleum capped by an impervious type of rock that prevents the oil from escaping to the surface. Oil fields are formed by the burial of organic matter. Most of the organic material exists as kerogen which is an insoluble precursor of crude oil. Kerogen is derived from organic matter by the metabolic activities of thermophilic anaerobic archaebacteria. Prescott (2005, page 889) states that microorganisms are present and prolific in oil reservoirs and other subsurface environments.

Oil shale is mudrock containing large amounts of organic matter. The mudrock may be true shale composed mainly of clay minerals, or it may be a carbonate rock such as dolomite or limestone. The organic matter in oil shale exists as kerogen derived from marine or freshwater algae, but also from planktonic organisms such as copepods and ostracods. Most of the kerogen in oil shale is not recognizable as organic remains.

Oil shale typically presents a fine lamination of alternating layers of minerals and organic matter indicating a succession of seasonal events involving sedimentation in quiet waters. Oil shales were deposited in large lake basins and in shallow seas. The black shales of the Cambrian period are marine oil shales.

Oil can be extracted from oil shale by heating it to 500 °C.

Thus, the deep burial of organic matter in conjunction with sediments has formed the world's fossil fuel reserves.

Conclusion

We have seen in this chapter that the sedimentary rocks of which the landmasses on Earth are composed fall mainly into three broad types:

- There are the micas and clays derived from the weathering – probably biological weathering by microorganisms – of basalts. Micas and clays have been preserved as metamorphic rocks in the form of schist and gneiss. Clays have been transformed into claystone, siltstone and mudstone. Mudstone has been compacted into shale, and shale may undergo metamorphism to become slate.

- There are the silica sedimentary rocks named quartz or chert. Extensive layers of chert exist as bedded chert, while isolated nodules called flints are found embedded in chalk. The shattering of quartz has produced sand which when cemented together by iron oxides or dissolved quartz becomes sandstone. There is also the secondary deposition of quartz by hydrothermally heated waters to form chalcedony veins.

- There are the carbonate sedimentary rocks in the form of limestone, chalk and dolomite. Limestone and dolomite may be hardened into marble by metamorphism.

One of the main physical processes contributing to the formation of landmasses is volcanism bringing molten magma from the mantle to the surface. Some mountain ranges have active or extinct volcanoes leaving basalt plugs, while high plateaus have been produced by flood basalts. Most volcanic activity occurring beneath landmasses, however, produces granite.

Molten magma rising up beneath sedimentary strata has been transformed into granite by the melting and mixing of bedded quartz. This enriches the magma with silica. The capping effect of the sedimentary strata prevents the granite from reaching the surface, and so it cools as massive intrusions below surface. These granite intrusions form the basement rock of mountain ranges on land. Granite tends to be exposed by weathering and erosion over time.

The redeposition of minerals is a major physical process contributing to the formation of sedimentary rock strata. The rocks of landmasses undergo constant weathering and erosion by ice and water. The eroded sediments and pieces of rock are generally carried away and redeposited elsewhere, becoming incorporated into new rock strata. Hydrothermally heated waters are also instrumental in the dissolving and redeposition of some minerals such as secondary quartz as chalcedony.

Another major physical process which transforms rocks is heat from volcanic activity combined with pressure from overlying layers of rock to produce metamorphic rocks. Metamorphic rocks tend to be harder and more resistant to weathering than the original rock, and so ancient rocks are often metamorphic rocks which have been preserved for longer.

The main thesis of this chapter is that the sedimentary rock strata of landmasses have an organic origin – the minerals of which they are composed were originally deposited by various forms of life.

Limestone was formed by green and red algae and foraminiferans living in the oceans, while chalk is composed of the coccoliths of

176

golden algae. Limestone reefs are built by corals which harbour algae, but also by sponges, mollusks and colonial Bryozoans. The limestone formed on the ocean floor has been uplifted to become land.

It is well-known that radiolarite chert was formed from the silica tests of radiolarian protozoans. I propose that bedded chert – which formerly may have been very extensive – was formed by diatoms living as plankton in the oceans. Diatoms are motile golden-brown and yellow-green algae and there are other silicoflagellates. If this were so, then pure quartz (SiO_2) is of biological origin. Quartz is resistant to weathering, but prone to shattering; for this reason most beds of chert would have been broken up by crustal movements, leaving the quartz as grains of sand now incorporated into sandstones or as a component of other rocks.

The earliest deposition of limestone and bedded chert on sea floors appears to coincide with the advent of protists 2.5 thousand million years ago. The protists include foraminifera, algae, diatoms and radiolaria which secrete tiny shells or tests for support and protection formed of calcium carbonate or silica. The motile, photosynthetic protists lived as huge blooms of plankton in surface waters, while protozoan scavengers lived on the seabed. The extensive layered deposits were formed from seasonal die-offs of planktonic blooms with the tests raining down to settle on the seabed.

I include the chert of banded iron formations in the same category as bedded chert. Banded iron formations date from 2.5 thousand million years to 1.8 thousand million years ago. According to this logic, the aerobic atmosphere of the Earth was mainly formed by photosynthetic algal protists, rather than by cyanobacteria.

Once the Earth had an atmosphere containing oxygen, aerobic bacteria could eek out a living upon basalt. Lithotrophic bacteria can obtain energy by oxidizing the reduced compounds contained in basalt. Communities of iron-oxidizing and manganese-oxidizing bacteria have been cultured from the weathered basalts of volcanically active areas of the ocean floor. These bacterial lithotrophs appear to be the principal producers of ferromanganese nodules and crusts found on the ocean floor, and to be active in the initial decomposition of basalt into the secondary minerals of mica and clay.

Clay contains the elements vital for life. Plants obtain mineral elements from clay. Clay is also the substrate for lower invertebrates such as marine worms. The clay sediments of the ocean floor contain microscopic worms and other organisms which become the organic matter content of sediments. The sediments are sifted for organic matter by larger deposit-feeding marine worms such as Polychaete Tube worms and Hemichordate Acorn worms which may be 3 metres long. These worms are known to play an important role in turning over the bottom sediments of the oceans.

I propose that the basalt transformed into mica and clay by aerobic lithotrophic bacteria, may then be transformed into mud containing organic matter by deposit-feeding marine worms. Mud is compacted into shale, and shale may be metamorphosed into slate.

Thus, it appears that the sedimentary rock strata of the Earth were formed under aerobic conditions by aerobic forms of life. The emphasis must be on aerobic conditions since it is only under these conditions that solids are produced rather than the gases, water and the soluble reduced compounds of anaerobic conditions.

The first types of sedimentary rocks were formed as the seabed. Later processes raised these sedimentary strata above sea level to form land. This is the subject of the next chapter, Chapter 21.

The protists of the Precambrian transformed the face of the Earth; they were the first to produce rocks unique to planet Earth – the sedimentary rock strata and granite of Earth's land masses.

Bibliography

Atlas, Ronald M., R. Bartha (1981) *Microbial Ecology: Fundamentals and Applications* Addison-Wesley Publishing Company

Cloud, Preston E. (1973) Paleoecological significance of the banded iron formation. *Econ. Geol.* Vol. 68, pages 1135-1143.

Cloud, Preston E. (1988) *Oasis in Space: Earth History from the Beginning* W.W. Norton & Company, New York

Conway Morris, Simon (1993) The Fossil Record and the Early Evolution of the Metazoa. *Nature* Vol. 313, pages 219-225.

Conway Morris, Simon (1998) *The Crucible of Creation: the Burgess Shale and the Rise of Animals* Oxford University Press.

Deer, W. A. et al. (1966) An Introduction to the Rock Forming Minerals Longman

Encyclopedia Britannica 2011 Standard Edition: Basalt; Bentonite; Chert; Copper; Feldspar; Gneiss; Gold; Granite; Loess; Mineral; Oil shale; Quartz; Schist; Sedimentary rock; Silica mineral; Silver; Stromatolite.

Encyclopedia of Mineralogy 2nd edition (1990) Roberts, W.L., T.J. Campbell, G.R. Rapp Jr.

Grzimek's Animal Life Encyclopedia 2nd edition Volumes 1-2 (2004) Michael Hutchins (Series editor) Thomas Gale

Han, T.M. & B. Runnegar (1992) Megascopic eukaryotic algae from the 2.1-billion-year-old Negaunee iron-formation, Michigan. *Science* Vol. 257, pages 232-235.

Ince, Martin (2007) *The Rough Guide to the Earth* Rough Guides Penguin Group

Lamb, Simon & David Sington (1998) *Earth Story: The Shaping of Our World.* BBC Books

Mader, Sylvia S. (2007) *Biology* Ninth Edition McGraw-Hill Companies Inc.

Margulis, Lynn & Karlene V. Schwartz (1982) *Five Kingdoms: an Illustrated Guide to the Phlya of Life on Earth.* 3rd edition. W. H. Freeman

Prescott, Lansing M., J.P. Harley, and D.A. Klein (2005) *Microbiology* 6th edition McGraw-Hill

Schopf, J.W. (1968) Microflora of the Bitter Springs Formation, Late Precambrian, central Australia. *Journal of Paleontology* Vol. 42, pages 651-688.

Shapiro, Robert (1986) *Origins: A Skeptic's Guide to the Creation of Life on Earth* Heinemann London

Templeton, A.S., Staudigel, H. and Tebo, B.M. (2005) Diverse Mn(II)-oxidizing bacteria isolated from submarine basalts at Loihi Seamount. *Geomicrobiology Journal* Vol. 22, pages 129-137.

Tunnicliffe, Verena (1992) Hydrothermal-Vent Communities of the Deep Sea. *American Scientist* Vol. 80, pages 342-?

www.ucmp.berkeley.edu/precambrian/bittersprings.html

CHAPTER 21

THE MECHANICS OF THE EARTH

Introduction

My aim in this chapter is to present a hypothesis that explains the causes of the geological processes observed on the face of the Earth. I offer my own theories for examination despite the popularity of current theories.

In this chapter I call into question the theories of Continental Drift, Plate Tectonics, Plume Theory or the Hot Spot Hypothesis, the Sea Floor Spreading Model and Ocean basin subsidence.

Plate Tectonics is a surprisingly emotive subject. Any break with orthodoxy seems to be highly risky. However, I have decided that it is worth the risk.

The important thing to note with this subject – Earth sciences – is that many 'facts' are theory-dependent; if the theory turned out not to be a true description of the geological processes occurring on Earth, then many so-called facts would evaporate away with the theory. Therefore, it is as well to be wary of theory-dependent facts in Earth sciences.

1. Land formation

Older geological theories give no explanation for the existence of land since it was assumed to have always been there, and therefore did not demand an explanation.

We now know that other planets in the solar system, where observation has been possible, do not have landmasses. They have surfaces consisting of basalt rocks. The ocean basins on Earth also consist of basalt rocks. If these basins were not filled with liquid water, their surface would resemble that of other terrestrial planets. Thus, the continents formed mainly of granite and sedimentary rock strata do not have counterparts on other planets.

According to the view of Earth history that I am proposing here, there were no continental landmasses on Earth during the long Precambrian period between 3800 and 600 million years ago. In the beginning there were no oceans; water started to accumulate on the face of the Earth during the Precambrian forming pools which later became shallow seas. Volcanoes projected above the waters of shallow seas as volcanic islands formed from basalt.

The formation of the first sedimentary rocks on the seabed was linked to the advent of an aerobic atmosphere. Three thousand five hundred million years ago cyanobacteria started to release free oxygen into the atmosphere. This allowed aerobic bacteria to perform the first step in the creation of the rocks of Earth's crust by transforming basalts into secondary minerals. Photosynthetic single-cell algae contributed to oxygen in the atmosphere allowing other protists to flourish in the seas as plankton and bottom dwellers. Algal and protozoan protists deposited sediments on the seabed in the form of their tiny shells. In this way sedimentary strata started to be formed at the bottom of shallow seas from about 2500 million years ago.

Starting 570 million years ago the Palaeozoic era with its multicellular life forms saw the building of limestone reefs by corals and other reef-builders in shallow seas. Around this time marine worms started forming mud on the seabed. Muddy sediments later became mudstones compacted into shale.

Land appeared out of the waters by the Ordovician period of the Palaeozoic era 500 million years ago and was colonized by the first primitive plants. The strata of sedimentary rocks laid down in shallow seas appear to have been raised up to form landmasses by faulting and folding of the marine rock strata. This first land was low-lying and swampy. It became a habitat of swamp forests with giant spore-bearing trees.

At the end of the Permian period and the close of the Palaeozoic era 250 million years ago land was further raised up by volcanic activity. Magma injected into the base of continental landmasses formed granite intrusions which became the core of mountain ranges. The higher ground now experienced coldness in some regions and dry conditions in others.

The globally warm conditions of the Palaeozoic era appear to have prevailed throughout the Mesozoic era of the dinosaurs and through the Palaeocene and Eocene epochs of the Cenozoic era until about 35 million years ago. The Oligocene epoch 35 million years ago marks the beginning of the present ice age whose effects are noted at the North and South Poles. The Oligocene saw the spread of grasslands in temperate regions with a fluctuating pattern of warm and cold phases in the climate of these regions. Thus, the Earth's climate appears to have been warmer for most of Earth history, but it is now in a cooler phase.

Over the past 500 million years landmasses have been built up by the geological processes of faulting, folding and volcanism. The onset of the present ice age 35 million years ago has led to the destruction of land through erosion especially at high latitudes.

In former times there was abundant life in the seas and on land at the North and South Poles showing that warmer conditions prevailed in previous eras. This may be partly due to the Earth's atmosphere formerly having more carbon dioxide in its composition – carbon dioxide that has now been sunk into limestone rocks and incorporated into living organisms inhabiting the face of the Earth – the warmer conditions of former times may also be due to cycles in the Earth's orbit and tilt on its axis affecting seasonality in temperate zones.

Another factor which may have caused the onset of the present ice age 35 million years ago which brought permanent ice to the poles may be the amount of water on Earth; the depth of the oceans would affect ocean currents and the heat exchange that occurs between waters in tropical seas and in polar seas. The shallow seas of the Palaeozoic era may have been a factor in warmer conditions at the poles, while the deep oceans that now cover the Earth may be linked to colder conditions.

Landmasses appear to have been more extensive in former times. Under the current cold conditions at the poles and in temperate regions land destruction is occurring. Especially in cool climates erosion of the land by water is a dominant factor. Water erosion takes the form of rain dissolving rocks, rivers carving river valleys and floods creating flood plains. Ice breaks apart rocks, and glaciers grind their way through river valleys gouging out the landscape on their way to the sea. Waves along coastlines pound away cliffs, leaving rock platforms beyond beaches of pebbles and sand.

The result of this is that older landmasses have been eroded away in temperate and cold regions. The islands of tropical seas are often of volcanic origin, while the islands found in temperate and polar seas tend to be pieces of a landmass that has been eroded away. The erosion process leaves chunks of resistant rock stranded out to sea as islands.

At one time the continents were much larger than they are now and the oceans were much shallower. The extent of continental shelves is an indication of the former extent of continents. Continental shelves have water only 100-200 metres deep. Some continental shelves are only a few tens of kilometres wide but many extend to hundreds of kilometres from the coast. The average width of continental shelves is 65 km. At the edge of the continental shelf there is a plunge down to the abyssal plains thousands of metres below – this is the true ocean floor. The continental shelves have been formed by wave erosion of the land and they constitute a zone of deposition of sediments from land erosion.

Continental shelves show that the landmasses of the Earth were once joined to each other. The Arctic had a landmass during the Palaeozoic era now eroded away. The American continent was connected to Eurasia between Canada, Greenland and an Arctic continent as well as between Alaska and eastern Siberia; South America was connected to North America via Central America; Antarctica was joined to South America and New Zealand; Australia was joined to New Guinea and some of Indonesia; Malaysia was joined to Asia; and Europe was joined to Africa.

Deep oceans are a comparatively recent feature of planet Earth. The average depth of oceans today is 4000 metres. I propose that it is the rise of sea level over geological time that has led to the inundation of low-lying land. The continents were once all joined via continental shelves that are now under water.

I have given a general description of the formation of landmasses and the effects of climate changes over geological history. The early landmasses, although unlike the present landmasses in terms of mountain ranges, coastlines and climate, would still have been located in the same relative positions. This understanding of landmasses does not include the notion that landmasses have moved around the face of the Earth.

2. Continental Drift Theory

The currently popular view to explain the form and present distribution of the continents on the face of the Earth comes under the title of Continental Drift. It is an old theory that had been rejected by the majority of geologists until it was resurrected again in support of the Theory of Plate Tectonics.

The Theory of Continental Drift was proposed by Alfred Wegener (1880-1930), a German polar researcher in 1915. He proposed that in the Palaeozoic era (before 225 million years ago) all the continents were joined together into one enormous supercontinent which he called Pangaea.

Wegener thought, as many others did, that the Earth's mantle was formed of liquid molten rock and he conceived the idea that the continents floated upon the mantle like icebergs floating on the sea. Wegener proposed that at the end of the Palaeozoic era, the supercontinent Pangaea broke up and fragments of it drifted to their present positions. The fragments became the present-day continents while the gaps in between became the oceans.

The majority of geologists rejected Continental Drift and physicists opposed the theory since they declared the lateral motion of continents to be impossible. Thus, the Theory of Continental Drift was held in little regard until it was revived by Sir Edward Bullard at Cambridge University in 1965. Plate Tectonics Theory formulated in the 1960s represents the surface of the Earth as being formed of lithosphere plates some 50-100 km thick which move by being constantly formed and destroyed. The plates are said to carry the continents upon them (the continents have a thickness on average of 35 km). Hence the idea that the continents have drifted ties in with the idea that the plates move over the surface of the Earth.

There are three lines of evidence presently used to support the notion of Continental Drift:

- The shape of the eastern coastline of South America can be matched to the western coastline of Africa. Thus, they are said to have at one time been joined.

- There are great similarities between ancient floras and faunas on continents that are now separated. For example, the fossils of the seed fern *Glossopteris* which flourished 270 million years ago are found in South America, southern Africa, India and Australia.

- The magnetic axis conserved in rocks (palaeomagnetism) has been used to determine how the continents fitted together to form the supercontinent and the direction they took when they drifted apart.

The theory in its present form, presents all the continental landmasses as having been amassed together 120 million years ago (this corresponds to the Cretaceous period of the Mesozoic era).

The Theory of Continental Drift is now widely accepted, but not by me. These are the reasons for my non-acceptance of the theory:

Wegener presented continents as floating on a mantle of molten magma. The mantle, however, is not composed of molten magma; it is mainly composed of solid rock with only isolated magma chambers. Thus, the likening of the mantle to the sea is not particularly enlightening. I do not believe that landmasses move relative to each other; they are not like icebergs floating on the sea.

I think that the matching of continental coastlines like badly-fitting pieces of a jigsaw puzzle is a very crude way of doing science. The exercise is more akin to being able to see a face in a rock if you look hard enough.

I see no reason why land should have formed in only one place on the surface of the Earth as a supercontinent. The Theory of Continental Drift gives no explanation for the formation of Pangaea in the first place. It seems that it is just assumed that land has always been there. The processes which have actually formed land, in my opinion, have occurred all over the globe, not just in one place.

I think that the maps constructed from palaeomagnetism are based on theory-dependent data – if the theory collapses this type of data disappears with it.

I am only impressed by the distribution of ancient floras and faunas as a line of evidence. However, I believe that the discovery of similar ancient species of plants and animals on widely separated continents can be explained by former connections between the continents and very different climatic conditions in the past. The former existence of land bridges between continents that were greater in extent can better explain the migration of animals and the colonization of plants.

3. Shrinkage and the Earth's crust

We have seen how the sedimentary strata which form a major part of the Earth's crust have been laid down by forms of life and physical processes upon the face of the Earth. The Earth's crust forms a thin, cool jacket around the Earth only a few kilometres thick compared to the 2900 km thickness of the mantle.

In Chapter 18 it was noted that the molten hot core of the Earth, although insulated by the mantle and crust, is losing heat to the outside and cooling down. When iron is heated up it expands, and when it cools down it contracts (see Appendix for this chapter for further details on the shrinkage of iron). It is a logical deduction that the iron core of the Earth, as it cools down is shrinking in size. It occurred to me upon reflection that while the interior of the Earth must be shrinking as it cools, its outer crust is not shrinking – the outer crust is cold in comparison with the core, and the sedimentary strata are hard and brittle. Over time, as the Earth's core shrinks, the hot mantle may adjust itself while the cold outer crust would not be able to shrink and tensions would be built up. These tensions would be the cause of folding, faulting and subducting of the Earth's crust.

The **Earth Shrinkage Hypothesis** can be formulated thus:

The Earth's core is cooling and shrinking in size, while the Earth's crust is essentially cold and resistant, and does not shrink. The Earth's hot mantle may adjust its size, while the crust is put under forces of compression. The crust adjusts to the smaller size of the interior regions of the Earth by the geological processes of folding, faulting and subduction.

After I had formulated this idea in the 1990s I found a similar idea in Alan Broms *Our Emerging Universe* (1961, page 164) [18]. Broms writes that shrinking produced by radioactive heating followed by cooling and cyclical ice ages causes wrinkling in the Earth's crust. He writes that continents formed of recent sedimentary strata are incompletely consolidated and structurally weaker and so shrinking produces mountain ranges. It also produces ocean trenches. The wrinkles are both north-south and east-west because shrinking reduces the Earth's circumference in all directions.

My idea is different from Broms' idea in that he attributes shrinking to the Earth's crust, while I attribute it to the Earth's core. He believes that there is a cyclical heating and cooling of the Earth's crust, while I think the key is the progressive cooling of the Earth's core. Broms sees mountain-building as being produced by shrinking and wrinkling of the Earth's crust, while I see mountain-building as being produced by the Earth's crust's *inability* to shrink. The inability of the crust to shrink, in my opinion, is the cause of wrinkling and the geological processes that we observe on the face of the Earth.

I see the Earth as being somewhat like a wizened apple – yes, even old fruit left in the fruit bowl can be a source of inspiration – as the inside of the apple dries up and shrinks, the skin of the apple which cannot shrink, goes wrinkly.

The crust can also be seen as forming an over-sized overcoat around the Earth's body.

---

[18] Broms, Alan (1961) *Our Emerging Universe* Laurel Editions Dell Publishing Co. USA

I believe that it is the inflexible nature of the Earth's crust that gives it its features. As the Earth's core shrinks and the mantle reacts, a gap would open up between the interior and the over-sized outer crust. A gap cannot be sustained, so various things happen to adjust the size of the outer crust to the size of the core and mantle. The sedimentary strata of continents fold into hills and mountains; at fault lines the crust breaks and one piece overrides another piece which slides beneath; and oceanic crust dives down at ocean trenches and disappears into the mantle. Folds, splits and tucks in the Earth's crust adjust its size. The tensions produced by shrinkage have caused the continents to be pushed up above the oceans which occupy the basins surrounding them.

These are the geological processes caused by Earth shrinkage:

Folding

Where sedimentary strata have a certain malleability possibly due to heat from volcanic sources, the tensions set up may cause the strata to fold. Folding is a way of reducing the size of the Earth's crust to adjust to the shrinking interior. Many mountain ranges have been formed by folding.

The sedimentary strata of continental crust have been extensively folded, and continents have been generally pushed up. There were seas within what are now continents where life forms laid down sediments. Marine sediments deposited in the ancient Tethys Ocean 55 million years ago, are now found high in the Himalaya Mountains and the southern part of the Tibetan plateau. They have been pushed up by folding and faulting.

Today's highest mountains are geologically recent. They have all been formed during the Tertiary period of the present Cenozoic era. The Himalayas have been built up during the past 65 million years. The Rockies have undergone phases of orogeny between 65 and 35 million years ago. The Andes have arisen within the past 15 to 6 million years.

Lower mountain ranges include the Atlas, Alps and Caucasus mountains. The Atlas Mountains underwent extensive folding during the Jurassic and Cretaceous periods of the Mesozoic era.

The Alps have been formed by folding. There have been folding episodes from about 40 million to 5 million years ago. Nappes are folded-over bodies of rock. The Alps cover less than 200 km from south to north – the direction of most of the over thrusting that forms the nappes. However, measurement of the nappes shows that folding has used up about 200 km of crust (Ince 2007, pages 70-71). Thus, the surface area has been greatly reduced.

Faulting
Where sedimentary strata are very hard and brittle, faulting is much more likely to occur than folding as the crust adjusts in size. Rocks are brittle when they are cool – that is with a temperature below 350 °C. In these areas tension will eventually lead to sudden ripping of the brittle crustal rocks and movement along a fault line.

Faults are another mountain-building geological process. Sedimentary strata originally laid down horizontally can be up-ended along fault lines such that the strata are now inclined at sharp angles or even almost vertical.

At thrust faults one piece of crust may be suddenly thrust on top of another piece of crust. In 1964 200 km of the Alaskan coastline moved on a thrust fault. The leading edge of Alaskan crust pushed upwards about 20 metres over the crust of the Pacific Ocean. The result was that this region rose out of the sea by 10 metres in some places while a wide area also subsided. The earthquake that occurred in Alaska in 1964 was found to be one of a recurring pattern of earthquakes that happen about every 800 years. Stranded coastal terraces on Middleton Island testify to previous movements along the same fault line accompanied by earthquakes (Lamb & Sington 1998, pages 68-72).

Fault rupture is felt as an earthquake. The earthquake focus which indicates the depth of the fault is often between 5 and 15 km under the surface. The rupture may propagate in one or both directions over the fault plane. When a barrier is met the rupture may cease or it may stop and recommence on the other side of the barrier or sometimes it may break the barrier and continue to rupture.

Faults may show the displacement of crustal rocks with one mass of rock rising above another mass, or masses of rock may move past each other sideways in strike-slip faults. At the time of a sudden slip, a fault may show relative displacements of slabs of rock of between a few centimetres and tens of metres. Movements of a metre often occur across fault lines during a single earthquake. Over geological time the relative displacements of rock masses may be in the order of hundreds of kilometres.

Mountain ranges are formed by multiple faults with thrust lines and they are zones of frequent earthquakes. The fault lines of America and the Andes Mountains run north-south, while the fault lines of the Himalayas run east-west.

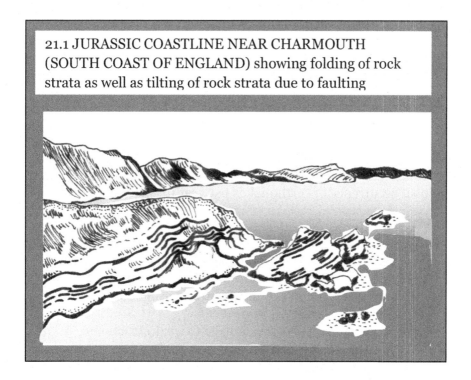

21.1 JURASSIC COASTLINE NEAR CHARMOUTH (SOUTH COAST OF ENGLAND) showing folding of rock strata as well as tilting of rock strata due to faulting

Earthquakes

Earthquakes occur when energy is suddenly released when masses of rock straining against one another suddenly fracture and slip. The fractures occur along fault lines. As a fault rupture progresses along or up a fault plane, rock masses are flung in opposite directions springing into a position where there is less strain. The sudden rupture of the

191

Earth's crust triggers vibrations in the surrounding rocks and the ground shakes with the passage of seismic waves. The greater the tension released along a fault line, the bigger the magnitude of the earthquake.

The most important earthquake belt affects the coastal regions around the Pacific Ocean. It has been estimated that 80 % of the energy presently released in earthquakes comes from epicentres in this belt which include New Zealand, New Guinea, Japan, the Aleutian Islands, Alaska, and the western coasts of North and South America. A second belt passes through the Mediterranean region eastward through Asia and the East Indies. The energy released in earthquakes in this belt represents about 15 % of the world total. There is also seismic activity along oceanic ridges such as those in the Arctic Ocean, the Atlantic Ocean, and the western Indian Ocean. The rift valleys of East Africa are another focus of seismic activity (*Encyclopedia Britannica 2011 Standard Edition*: Earthquake).

21.2 RICHTER SCALE FOR EARTHQUAKES

The Richter Scale is a measure of the magnitude of earthquakes. It goes from 1 to 8 or more. The following list is the Richter Scale and the global occurrence of earthquakes per year in each category:

- **1-2.9 micro** – generally not felt by people, though recorded on local instruments. Occurrence: more than 100 000 per year
- **3-3.9 minor** – felt by many people; no damage. Occurrence: 12 000-100 000 per year
- **4-4.9 light** – felt by all; minor breakage of objects. Occurrence: 2000-12 000 per year
- **5-5.9 moderate** – some damage to weak structures. Occurrence: 200-2000 per year
- **6-6.9 strong moderate** – damage in populated areas. Occurrence: 20-200 per year
- **7-7.9 major** – serious damage over large areas; loss of life. Occurrence: 3-20 per year
- **8 or higher, great** – severe destruction and loss of life over large areas. Occurrence: fewer than 3 per year

The San Andreas Fault is a major fracture in the Earth's crust running for 1300 km from the Gulf of California into the Pacific Ocean in the vicinity of San Francisco. Movement along the fault has been associated with occasional large earthquakes originating near the surface along its path. A disastrous earthquake occurred in San Francisco in 1906 and again in 1989, and a destructive quake occurred in Northridge in 1994 along a secondary fault line.

During the 1906 earthquake, the San Andreas Fault slipped along a segment 430 km long over land extending out to sea to an unknown distance. Along this line the ground was displaced horizontally by as much as 6.4 metres.

The largest earthquake of the 20th century occurred in Chile in 1960. The source of the earthquake was a fault displacement extending 1100 km along the southern Chilean coast with a major under thrusting of the Pacific Oceanic crust under the South American continent. The movement created a tsunami which caused destruction in Hawaii, Japan and along the West Coast of North America.

A reoccurrence of displacement along the same fault occurred on the 27th February 2010 at 3.34 am local time. The intense shaking which lasted three minutes was rated at 8.8 on the moment magnitude scale making it the sixth largest earthquake ever recorded by a seismograph. The epicentre of the earthquake was a few kilometres off the coast with the worst affected town being Concepcion, but the quake was felt in most of the south of Chile and earth tremors were felt as far away as Argentina and Peru. The earthquake triggered a tsunami which devastated several coastal towns in southern Chile and caused a wave to cross the Pacific Ocean although it caused little damage elsewhere.

Five hundred and twenty five people died in the 2010 earthquake in Chile. My children who were staying in San Antonio were among the people affected by the earthquake with no electricity or water and roads cut off. In 2011 on 2nd January an earthquake occurred in the same place rated at 7.1, then another earthquake occurred in February – I was in the north of Chile and felt the strong earth tremors. A fourth consecutive earthquake occurred in the same place on 2nd June 2011 rated at 6.2, followed by after-shocks rated at 5.2. The 2011 earthquakes in Chile caused less damage because all the buildings that were going to fall down had already fallen down in 2010.

Other major earthquakes occurred at about the same time as the Chilean earthquakes. Haiti was struck by an earthquake rated at 7.0 on the 12th January 2010 with 52 aftershocks. 316 000 people died in this earthquake and a million were made homeless. On 22nd February 2011 a major earthquake occurred in New Zealand rated at 6.3 in which 180 people died. On 11th March 2011 one of the most powerful earthquakes ever recorded hit Japan rated at 9.0 magnitude. It caused large scale damage and loss of life running to a death toll of 15 790. Many of the casualties were caused by a highly destructive tsunami following the earthquake.

Tsunamis
The sudden displacement of the seabed along a fault line during an earthquake or a submarine landslide arising from an earthquake may cause a sudden raising and lowering of a large body of water. This creates a wave a few metres high with a wavelength of hundreds of kilometres and a long interval between the arrival of successive crests in the order of tens of minutes.

When the tsunami reaches the shallow water of an inlet or harbour, the wave amplitude increases sometimes to 20 to 30 metres above mean sea level and the wave front may be nearly vertical. The speed of the water onrush may be 10 metres a second, hence the destructive potential. Before the first waves hit the coastline, there may be an extraordinary recession of water from the shore for up to half an hour before disaster strikes.

The most destructive tsunami ever recorded occurred on 26th December 2004 after an earthquake displaced the seabed off the coast of Sumatra, Indonesia. A series of tsunami waves flooded coasts from Indonesia to Sri Lanka, and washed ashore on the Horn of Africa.

The large number of earthquakes occurring each year some of which are followed by tsunamis shows the extent of the geological processes involving faults and subduction affecting the Earth's crust. The tremors may be palpable evidence that the Earth is shrinking and the crust readjusting itself.

Deep earthquakes

The occurrences of deep earthquakes at over 70 km depth have not yet been given a satisfactory explanation, but they may lend support to the Earth Shrinkage Hypothesis. If the interior of the Earth is contracting, one would expect adjustments to affect the mantle at all depths. Deep earthquakes relieve more stress than shallow earthquakes. The great Bolivia deep earthquake of 9th June 1994 with a magnitude of 8.3 occurred at a depth of 636 km [19].

Subduction

Deep oceanic trenches found around the margins of oceans reach down 6000-7000 metres below sea level. The deepest known trench is over 11 000 m deep. These deep narrow depressions are about 100 km wide and often filled with thick sediments. They always lie about 100 km offshore and parallel to chains of subduction volcanoes. It is here that oceanic crust slides into the Earth's interior. The slab of descending rock takes a gently inclined path, and then plunges into the Earth's mantle beneath volcanic arcs. Small, frequent earth tremors accompany the descent.

Seismological studies which identify earthquake foci indicate the path of the descending slab of rock. Along the periphery of the Pacific Ocean, earthquakes occur close to or landward of oceanic trenches at depths within the Earth of 55 km or less. With increased landward distance from the oceanic trenches, earthquakes occur at greater and greater depths to 500 km or more. Seismic foci define tabular zones approximately 20 km thick that dip landward at about 45 degrees beneath the continents (*Encyclopedia Britannica 2011 Standard Edition*: Deep-sea trench).

Subduction volcanoes represent 80 % of active volcanoes on Earth. Subduction zone volcanoes always lie 100-200 km vertically above the subducting crust. Subduction volcanoes may form mountain chains on land such as the Andes Mountains running the length of South America or arcs of volcanic islands such as the Aleutian Islands between Alaska and Asia. Island arcs often consist of a double row of islands and most are found around the Pacific basin. It appears that decompression of the mantle above the subducting slab occurs as subduction proceeds and this leads to volcanic activity.

---

[19] http://geology.about.com/od/earthquakes/a/aa_deeEQs.htm

Destructive earthquakes frequently occur at the site of island arcs. Most other areas in the world have shallow earthquakes, however, earthquakes associated with subduction have deep seismic foci located as much as 600 km below the base of the island arc (*Encyclopedia Britannica 2011 Standard Edition*: Island arc).

The subducting oceanic crust descends into the mantle laden with marine sediments containing organic matter. The heat of the mantle causes the combustion of the remains of dead marine organisms converting them into gases. The combustion of organic matter produces carbon dioxide, water vapour, sulphur dioxide and nitrogen and these gases are blown out of volcanoes. Subduction volcanoes contribute to the recycling of the remains of life back into atmospheric gases.

The Pacific Ocean crust is sliding beneath continents all around the Pacific Ocean margin. There are also ocean trenches in the north eastern part of the Indian Ocean and some small ones in the Atlantic Ocean. Elsewhere they are absent. Therefore, the main focus of subduction is around the Pacific Ocean.

Oceanic trenches where oceanic crust is subducted and disappears into the mantle appear to be as tucks in the Earth's crust reducing its overall size. This evidence supports the Earth Shrinkage Hypothesis.

Extraterrestrial support for the Earth Shrinkage Hypothesis

Folding, faulting and subduction are geological processes affecting the Earth's crust that can be explained by the notion that the Earth is shrinking in size. Earthquakes are produced by the movement of rock masses along fault lines and on a minor scale number over 100 000 a year. Tsunamis are produced by movement along fault lines under the oceans displacing large bodies of water.

These are some extraterrestrial lines of evidence in support of the hypothesis that the Earth is shrinking.

Mercury
Mercury shows evidence of shrinkage on its surface. Mercury has fault scarps that cut across the surface about 1000 metres high and several hundred kilometres across. Individual scarps often traverse different types of terrain cutting across old craters. According to Zeilik (2002, pages 179-180) the characteristics of the scarps imply that Mercury's radius has shrunk by 1 to 3 km since the scarps were formed. According to Taylor (1998, pages 165-166) the shrinkage of Mercury's radius has been between 2 and 4 km.

Taylor attributes the shrinkage of Mercury's surface to contraction of the mantle due to cooling and solidification at an early stage in its history. Zeilik attributes the shrinkage to cooling of the planet's core or surface. Zeilik likens it to the skin of a baked apple that wrinkles as it cools.

If there is evidence that Mercury has shrunk in size, this supports my contention that the Earth has shrunk in size. Having examined the matter closely and with a knowledge of apples, we know that the only way to produce wrinkles on an apple or scarps on a planet would be for the core to shrink while the skin or outer surface does not shrink, but instead becomes puckered.

Venus
Venus is akin to the Earth in size being only 5 % smaller and its density is almost the same as the Earth. The crust on Venus is folded and faulted in some regions by processes that resemble the mountain-building processes on Earth. There are rift valleys that cut the crust into sections. The wrinkling and puckering of Venus' surface may have been produced by shrinkage as I have explained it for Earth. Evidence for this on Venus is that some wrinkle ridges have been formed upon former ridged plains.

Recession of the Moon
It has been found that the Moon which is in a nearly circular orbit around Earth is receding away from the Earth by 3.82 cm a year (see footnote [20]). Various explanations have been offered to explain this.

---

[20] Ince 2007, page 40 states that the Moon is moving away from the Earth at 2.17 cm a year.

I would ask the question, is the Moon receding from the Earth, or is the circumference of the Earth shrinking away from the Moon? If the Earth were being reduced in size, the distance between the Earth and the Moon would be increasing.

4. Plate Tectonics Theory

Plate Tectonics is the name of the theory most favoured today to explain the geological processes seen on the face of the Earth. Plate Tectonics goes hand in hand with the Continental Drift Theory that I described above, with Plume Theory and with the Sea Floor Spreading Theory that will be explained shortly.

The Theory of Plate Tectonics proposes that the Earth's surface is divided into a number of large plates and several small plates. The large plates number between seven and twelve depending on which expert is consulted and they are thought to be moving relative to each other. The Theory proposes that the plates are composed of the lithosphere of the mantle with a thickness of 50-100 km – the lithosphere is the cooler outer part of the Earth's mantle – and that these lithosphere plates carry the Earth's crust.

The principal tectonic plates are named the Eurasian Plate, African Plate, North American Plate, South American Plate, Nazca Plate, Pacific Plate, Indo-Australian Plate and Antarctic Plate.

The continents are said to be less dense, and to float on the mantle, and so sit high on the Earth's surface forming dry land. (In reality the continents are implanted deeper into the mantle than oceanic floor, and therefore do not sit high upon it, but low into it). The theory proposes that as the plates move, the continents carried on the plates move with them.

The Theory of Plate Tectonics was formulated in 1967-68. There were many contributors to the body of this theory and it encompasses the older Theory of Continental Drift proposed by Alfred Wegener in 1915. The interactions at the boundaries of plates are thought to be responsible for volcanic activity, faults, earthquakes, and mountain-building processes such as folding. It must be noted, however, that

198

volcanoes and faults are also found in the middle of plates and this is not explained by the theory.

The plates are represented as being formed at mid-ocean ridges by the spreading of molten rock (mid-ocean ridges will be explained in the next section). As a plate is formed, an ocean basin forms. Rift valleys in continents are said to be places where plates also form. A rift valley is thought to be an incipient area of spreading that will one day form an ocean, splitting a continent apart.

Plate Tectonics asserts that the surface area of the Earth remains constant; the creation of new crust by volcanic activity along mid-ocean ridges being exactly balanced by the destruction of crust at subduction zones. In subduction zones cool rock is observed to slide back into the mantle at oceanic trenches. The idea for Plate Tectonics is that the subducted plate melts at a depth of between 300 and 700 km. It is then recycled by rising again as molten rock at mid-ocean ridges.

The driving force for Plate Tectonics is proposed as coming from convection in the mantle, although there is much dispute about this. The mantle is represented as having hot material rising at mid-ocean ridges, spreading horizontally at the surface, cooling and sinking back in at subduction zones. This is represented in diagrams as giant convection cells. The movement of plates is said to be made possible by the lubrication of water.

Plate Tectonics proclaims that plates move apart from the central rift valleys of oceans; they move past each other along transform faults; they are moving towards each other in places such as the Pacific margin; and they are colliding with each other and pushing up mountains in places such as the Himalayan Mountains. The moving plates are represented as carrying continents –they are thought to have caused the continents to split, regroup and drift apart again affecting the climatic conditions of the continents over time and the distribution of species. The Theory of Plate Tectonics is almost universally accepted. I, however, reject it. In the following discussions I give some of my reasons for rejecting the Theory of Plate Tectonics.

The Himalayas – colliding plates?
The 110 peaks of the Himalaya Mountains rise up between the low-lying Ganges Plain of India and the high Tibetan Plateau.

The Himalayas run east to west in four parallel ranges named the Outer Himalayas, Lesser Himalayas, Great Himalayas and Tethys Himalayas which extend into the Trans-Himalayas.

The story of the formation of the Himalaya Mountains and the concurrent disappearance of the Tethys Ocean is told with a view to demonstrating the Theory of Plate Tectonics and Continental Drift in all the current textbooks. The names used for landmasses and their locations reflect the explanations given by the theory. The story can still be told without a belief in Plate Tectonics or Continental Drift but the labels belonging to the theories must be dropped and some conceptualizations modified in accordance with the view adopted.

The formation of the Himalaya Mountain ranges is presented according to current thinking as the result of geological processes occurring at a plate boundary. It is envisaged, according to Plate Tectonics, that during the Mesozoic era some 180 million years ago the Indian-Australian Plate carrying the Indian subcontinent moved northwards towards the Eurasian Plate confining the Tethys Ocean in between the two landmasses. The landmass to the south is named Gondwana. A fragment of Gondwana, now India is thought to have then collided with Eurasia in the Tertiary Period 50 million years ago. When this occurred, the Indian subcontinental plate subducted beneath the Eurasian Plate at the Tethys trench. The plunging plate caused the seabed of the Tethys Ocean to be pushed up and it now forms the Plateau of Tibet. Thus, according to this view, the Tethys Ocean was eliminated by the convergence and collision of the Indian subcontinental plate with the Eurasian Plate.

According to the Earth Shrinkage Hypothesis, it could be envisaged that the Tethys Ocean existed within what is now the Eurasian landmass at the location of the present Himalaya Mountains and Tibetan Plateau extending through the Caucasus Mountains of Central Asia, the Alps, Carpathians and Pyrenees of Europe and Atlas Mountains of northwest Africa. The Mediterranean basin, Black Sea and Caspian Sea would be remnants of the once larger Tethys Ocean.

With the Earth Shrinkage scenario the Tethys Ocean basin would have been subjected to compression forces resulting from the imperative to adjust to a shrinking core. The evidence is that the seabed with its sedimentary strata underwent faulting (probably in four main east-west lines that now mark the four ranges of the Himalaya Mountains) and eventually buckled up to form high ground. The sedimentary strata which once formed the ocean floor underwent over-thrusting along fault lines and folding eliminating most of the ocean, except for some remnants.

Some of the Tethys ocean floor may have been subducted down an oceanic trench if one had existed. This would have resulted in subduction volcanoes, or this may not have happened at all. In any case, if a trench existed and crust was subducted down it, it would be oceanic crust and not continental crust, much less a large plate. (Plates carrying continents are not observed to disappear down oceanic trenches).

Whether one adheres to Plate Tectonics Theory or the Earth Shrinkage Hypothesis the geological processes which built the Himalaya Mountains can be viewed in almost the same way. The geological processes that built the Himalayas started with faults in the Tethys Ocean floor and the eruption of granite and basalt during the Mesozoic era some 180 million years ago. During the Cenozoic era the seabed of the Tethys Ocean was gradually pushed up and now forms the Plateau of Tibet. Over the past 30 million years the Himalayan Mountains were pushed up by strata of metamorphic rocks undergoing strong horizontal thrusting to form nappes.

Wave after wave of nappes were thrust southward over the Indian landmass for as far as 97 km (60 miles). Over time these nappes became folded, contracting the former surface area occupied by the Tethys Ocean by over 400 km (250 miles) horizontally. Some authorities estimate it at 800 km (500 miles). Rivers matched the rate of uplift by cutting into the mountains and eroding vast quantities of material which was carried to the Ganges Plain where it was dumped over the flood plain.

Over the past 600 000 years the Himalayas have become the highest mountains on Earth by intense uplift – along the northernmost nappes and beyond, granite intrusions have produced the staggering crests seen today (*Encyclopedia Britannica 2011 Standard Edition*:

Himalayas). Some peaks like Mount Everest have fossil-bearing Tethys sediments on their summits as a testimony to their origin as ocean floor.

As far as geological processes are concerned, the main difference in the alternative scenarios would be that with the Earth Shrinkage Hypothesis nappes would be considered to be over-riding layers of strata all moving from their positions in the north in a southwards direction. The result of the over thrusting of nappes is, it seems, a contraction of the surface area of the Earth's crust of between 400 and 800 km which accords well with the Shrinkage Hypothesis. The Plate Tectonics scenario presents the nappes as upper strata originating as part of the southern landmass and folding back over themselves at the point where the Indian subcontinental plate slid beneath the Eurasian Plate at the so-called Tethys trench. This would make the top strata older than the lower layers of strata belonging to the nappe.

If an Indian subcontinental plate were colliding with a Eurasian Plate, the forces produced could only account for the formation of the Himalaya Mountains adjacent to the Ganges Plain; this explanation could not be applied to the Alpine Mountain ranges and other mountain ranges of Europe, Eastern Europe and the Balkans since these mountains are located above the African landmass. The forces involved in the Earth Shrinkage Hypothesis could account for the more generalized uplift of the former Tethys Ocean floor into all of the mountain ranges belonging to the Alpine-Himalayan highlands.

Lithosphere and asthenosphere?
The Theory of Plate Tectonics requires that the mantle be divided into two parts; lithosphere is the name given to the outer part of the mantle which is cooler and thus thought to be firmer and which according to the theory is divided into plates. Below the lithosphere, the mantle is given the name asthenosphere which being more interior is hotter. According to the theory the asthenosphere is partially molten or behaves as if it were liquid or malleable because molten. If it were not liquid or malleable, the plates could not float upon it or move relative to it, and there could not be convection cells within it.

Seismic studies do not support this model of the Earth's mantle – the model of the Earth's mantle divided into lithosphere and asthenosphere is required by the Theory of Plate Tectonics or the theory breaks down.

As already touched upon in Chapter 18, section 1, seismic studies track the paths of vibrational waves produced by earthquakes. Earthquakes give out body waves which travel within the Earth and surface waves. The body waves consist of P and S waves. P stands for Primary waves; compression or longitudinal waves. P waves have a back and forth motion in the direction of propagation and travel at different speeds through solids, liquids and gases and with different hydrostatic pressures. The speed of P waves varies between 6 and 11 km per second. S stands for Secondary waves also called shear or transverse waves. S waves cause solid media to move back and forth perpendicular to the direction of propagation, shearing in one direction then the other. S waves travel at about 3.4 km per second at the surface to 7.2 km per second near the boundary of the core. S waves do not travel through liquids.

The absence of S waves in the vicinity of the Earth's core is a compelling argument for the core being liquid. I believe that the presence of S waves within the Earth's mantle is an equally compelling argument for the mantle being solid.

The Earth's mantle consists of silicate minerals which despite having high temperature which would make them molten at the pressure of the Earth's surface remain solid because they are under very high pressure produced by overlying layers of rock.

Seismic studies show that there are regions of liquid molten magma in certain parts of the solid upper mantle where active volcanoes occur. Molten magma chambers form under domes beneath crustal rocks when pressure is removed from the mantle and decompression occurs. Decompression of the mantle causes the silicate minerals of the mantle to melt and erupt upwards to form magma chambers. Magma chambers are evacuated at the surface via conduits which form volcanoes.

As an example, magma chambers have been detected beneath the East Pacific Rise by seismic experiments. The partially molten or molten rock of the chamber slows down the speed of seismic waves and strongly reflects them. The studies show that the depth of the top of the chambers is about 2 kilometres below the seafloor. Other studies suggest that the chambers are about 2-6 kilometres deep and maybe 1-4 kilometres wide. Magma chambers have been mapped along the trend of the mid-oceanic crest between 9° and 13° N latitude (*Encyclopedia Britannica 2011 Standard Edition*: Ocean). Other magma chambers that have been mentioned in Chapter 19 section 4 are one beneath Yellow Stone Park, Wyoming USA and another beneath the islands of Hawaii.

My conclusion is that the mantle is solid; it does not have continents or plates 'floating' upon it. Neither does the mantle have convection cells within it with molten rock churning round in cells the size of oceans – so-called convection cells are *only required* by the Theory of Plate Tectonics to drive the movement of plates.

Plume Theory
There are four main types of volcano:

- Subduction volcanoes on land and island arc volcanoes in oceans are associated with subduction zones.

- Volcanoes forming part of mountainous areas in the middle of continents. These may be shield volcanoes or plateaus which do not resemble volcanoes known as Large Igneous Provinces.

- Mid-oceanic ridge systems found below sea level in the oceans, but above sea level in Iceland.

- Volcanic islands found as groups and archipelagos.

According to Plate Tectonics Theory volcanism is a phenomenon associated with tectonic plate boundaries; however, a significant number (5 %) of known volcanoes occur in the middle of plates. The islands of Hawaii represent the biggest active volcanic system on Earth and Hawaii is located in the middle of the Pacific Plate. Yellowstone Park with its active volcanism is located in the middle of the North American Plate. There are also the active volcanoes of the East African Rift Valley found within the African Plate.

In order to explain the occurrence of 'intraplate volcanoes' the Theory of Plate Tectonics recurs to another hypothesis –Plume Theory or the Hot Spot Hypothesis.

Plume Theory or the Hot Spot Hypothesis is used as an explanation for the volcanic islands of Hawaii found in the Pacific Ocean where no plate boundary is to be found. 'Intraplate volcanoes' are called, in accordance with current theory, hot spot volcanoes.

Plume Theory presents the idea that plumes of hot magma rise from close to the Earth's molten core through the mantle to the surface by slow convection. This creates hot spots on the Earth's surface. The hot plume in the mantle melts the crustal rocks above and bursts out periodically onto the Earth's surface. The hot spot remains stationary, while the lithosphere plate such as the Pacific Plate moves crustal rocks over it. As a lithosphere plate moves over a hot spot, a chain of volcanoes are formed in sequence. The hot spot is depicted as acting like a blow torch switched on and off, creating volcanoes.

The evidence used to support Plume Theory is that the islands belonging to Hawaii get steadily younger from northwest to southeast. Active volcanism is only seen on Big Island which is the most south-easterly of the islands. To the northwest, Kaui is the last island and the oldest at 5 million years old. Beyond it are a string of seamounts which are old islands that have been eroded away and now exist below sea level.

Plume Theory or the Hot Spot Hypothesis remained popular until it became possible to make seismic images of the deep Earth. The new technique did not detect deep plumes emanating from deep within the mantle below hot spot sites; only shallow plumes were detected.

The Earth Shrinkage Hypothesis does not associate volcanism with any sort of 'plate boundaries' or hot plumes of magma emanating from near the Earth's core making hot spots in the middle of plates. Evidence suggests that volcanism is a phenomenon produced by decompression of the mantle for which there are a number of causes. Globally, the Earth Shrinkage Hypothesis presents the view that volcanism is associated with mantle adjustment to a shrinking core.

According to the Earth Shrinkage Hypothesis, as the Earth's core cools and shrinks, the interior of the Earth shrinks in size and the Earth's crust must also reduce its size. The crust reduces in size by folding up into mountains, faulting with one slab of crust overriding another slab of crust, and forming oceanic trenches that tuck oceanic crust into the mantle. The Earth's mantle in between the shrinking core and reducing crust must also adjust in size; it does so by volcanism. Decompression in parts of the upper mantle forms magma chambers that are emptied via volcanic eruptions. This overflow of molten rock onto the Earth's surface moves material from the mantle to the surface allowing the mantle to adjust to the pressures upon it and undergo a reduction in size.

When a magma chamber forms beneath ocean floor, the resultant volcanoes will form an archipelago of volcanic islands. When a magma chamber forms beneath a continent, the escaping molten magma may form granite intrusions by mixing with layers of sedimentary rock and solidifying below the surface. Mountain ranges on land may be formed of a combination of large masses of granite and some volcanic peaks where volcanic ash has erupted at the surface. This widespread volcanic activity has not been associated with any 'plate boundaries'.

Movement of tectonic plates?
The speed at which plates move relative to each other can be measured by the global positioning system (GPS). The plates are moving relative to each other at speeds of on average less than 10 centimetres a year, but up to 20 cm a year. The floor of the eastern Pacific Ocean is moving towards South America at about 9 cm per year. During the last 10 million years, the Pacific Ocean crust has slid under the western margin of South America and sunk nearly 1000 km into the Earth's interior (Lamb & Sington 1998, page 80). It is claimed that on average the continents are moving at 2 cm a year.

I certainly believe that the Earth's crust is moving and this is what the Earth Shrinkage Hypothesis would predict. That parts of the Earth's crust are moving together such as along the Pacific Ocean margin and that in other areas the crust is being crumpled and pushed up such as in the region of the Himalaya Mountains accords well with the Earth Shrinkage Hypothesis since this is what one would expect if the Earth was becoming smaller.

I do not believe that the Earth's crust is divided into a specific number of plates carried upon pieces of mantle lithosphere as described by the Plate Tectonics Theory. I do not believe that these so-called plates are moving across the planet.

I certainly accept GPS monitoring which shows that various parts of the Earth's surface are moving at various speeds. I would, however, ask, what is the overall effect of the individual movements and the direction of movement? The moving together of parts of the crust in some places and buckling in other places may produce, if calculated, the overall effect of the Earth being reduced in size.

5. Sea Floor Spreading Model

The Sea Floor Spreading Model is a crucial support to Plate Tectonics Theory since it aims to explain how the plates are formed. The Sea Floor Spreading Model was suggested by the American Harry Hess in the early 1960s and the Canadian Tuzo Wilson in 1965.

Mid-ocean ridges

Let us recall that mid-ocean ridge systems are broad zones of shallower water that mark the location of underwater mountain ranges. Mid-ocean ridge systems have a series of elongated ridges some 1000 to 1500 metres high. The crests of the ridges tend to be rugged and there is often a longitudinal rift valley along the top of the ridge from which emanate fresh lava flows. The ridges have high heat flow and shallow earthquakes. The mid-Atlantic ridge system has an elevation of 2500 metres and covers an area 1000 km wide.

The basalt of oceanic crust comes from the upwelling of the mantle at ocean ridges through long fissures. Along mid-ocean ridges the Earth's crust is very thin; it is only 1–4 km in thickness. Chambers of magma lie beneath the surface. The volcanic activity of the ridges seems to be associated with magma from superficial upwelling of the upper mantle.

Mid-ocean ridges nearly girdle the planet and they are nearly always hidden beneath the sea. However, the mountains of central Oman are peaks that were formed as deep sea floor and became pushed up as land. The Mid-Atlantic Ridge system is also seen above sea level in Iceland. The island of Iceland has constant volcanic activity and earthquakes. During volcanic eruptions long fissures open up, releasing molten basalt.

Mid-ocean ridge systems are accompanied by extensive faulting at right angles to the ridges called transform faults. Transform faults have allowed slabs of crust to move past each other on either side of the ridges.

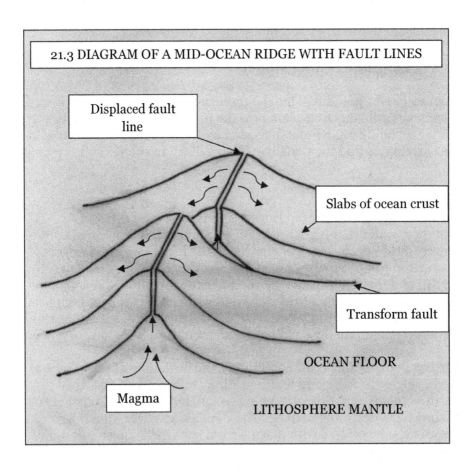

21.3 DIAGRAM OF A MID-OCEAN RIDGE WITH FAULT LINES

Displaced fault line

Slabs of ocean crust

Transform fault

OCEAN FLOOR

Magma

LITHOSPHERE MANTLE

Sea Floor Spreading
The idea of Sea Floor Spreading is that mid-ocean ridges create new oceanic crust. According to this theory the mid-oceanic ridge systems represent rift systems where two new plates of lithosphere and basalt ocean crust form and move apart from a central rift valley. There is an upwelling of the mantle at these 'seams' that run through the middle of oceanic basins. In this way new ocean floor is thought to spread either side of the ridge system.

According to the theory, the Earth's surface area does not increase by this process because the ocean crust and lithosphere are subducted at ocean trenches. At subduction zones two plates meet, and one plate dives beneath the other carrying lithosphere mantle and crustal rocks down into the asthenosphere mantle. It is thought that at some 300-700 km below surface the plate is recycled by becoming part of a convection cell within the mantle. Therefore, supporters of the Sea Floor Spreading Model propose that the Earth's surface is formed of plates which are continually formed at mid-oceanic ridges and continually destroyed at oceanic subduction trenches; this is associated with the notion that the Earth's processes are always balanced and that the Earth always remains the same size.

The Sea Floor Spreading Model is relatively simple and straight forward, unlike the reality of ocean floors.

Mid-oceanic ridges are not always found in the middle of oceans. The East Pacific Rise of the Pacific Ocean leads into the San Andreas Fault of the Gulf of California of the North American continent, while the Carlsberg Ridge which is part of the Mid-Indian Ridge of the Indian Ocean leads into the Red Sea of North Africa.

Mid-ocean ridge systems do not have a single line of volcanism along a central rift valley as the Sea Floor Spreading Model requires. Ince (2007, page 63) writes that the mid-Atlantic ridge is active across some undefined width; there is not just a narrow line of active volcanism. The mid-Atlantic ridge is seen above sea level in Iceland. On this island 70 volcanoes have erupted in the past 10 000 years. Effusions of lava from long, parallel fissures known as fissure vents can be seen in Iceland. Fissure vents erupt in sequence with new eruptions occurring a few hundred to a few thousand metres from the earlier fissure eruption. This outpouring of magma has built up thick lava plateaus in Iceland. The Laki fissure erupted in 1783 producing a crater row 25 km long, and an area of lava flows of 565 square kilometres (*Encyclopedia Britannica 2011 Standard Edition*: Volcano).

The Earth has 20 major oceanic trenches and 17 of these are found around the margins of the Pacific Ocean. The Indian Ocean has one major trench, the Java Trench close to Indonesia, while the Atlantic Ocean only has two small trenches; the Puerto Rico Trench north of the Caribbean islands and the South Sandwich Trench east of Drake Passage between South America and Antarctica. Thus, oceanic trenches are absent around many ocean margins.

The lack of a mid-ocean ridge system in the middle of the Pacific makes it difficult to see how large parts of the Pacific Ocean floor could have been produced if ocean floor forms at mid-ocean ridges. The many ocean trenches around the Pacific Ocean show, however, how ocean floor may be disappearing down trenches around the ocean margin. The Atlantic Ocean presents a clear mid-oceanic ridge system running down the middle – the Mid-Atlantic Ridge from which it is thought to have formed, but the trenches where oceanic crust is supposed to disappear are mostly absent. The Indian Ocean presents a mixed picture of a mid-oceanic ridge system and just one large trench. It is difficult to see how the Earth would maintain exact balance in its processes, given its topography.

According to Plate Tectonics plates are between 50 and 100 km thick and consist of lithosphere mantle and crust. Seismic studies have shown that the magma chambers from which fresh basalt lava emerges at mid-ocean ridges are 2 km below the surface, and probably 2-6 km deep and 1-4 km wide. If tectonic plates are formed at central mid-ocean ridges, how would a magma chamber 2-6 km deep produce a plate 50-100 km thick?

According to Plate Tectonics Theory plates are destroyed at oceanic trenches. How would a plate 50-100 km thick disappear down a trench 6-7 km deep? There seems to be some disproportion here.

I suggest that only basalt crust is produced at mid-ocean ridges by magma oozing onto the ocean floor and only ocean crust disappears down oceanic trenches as a slab of rock, rather than as a 'tectonic plate'. Oceanic crust has an average thickness of 7 km, consequently, its disappearance down oceanic trenches seems possible.

Seismological studies show that the descending slab of rock at a subduction zone is about 20 km thick and descends at 45° beneath the continent. The data from earthquakes has been interpreted as showing the descent of a lithospheric plate with its associated crust into the asthenosphere mantle, with oceanic trenches being the topographic expression of this movement (*Encyclopedia Britannica 2011 Standard Edition*: Deep-sea trench). I suggest that these seismological studies show the descent of a slab of oceanic crust into the mantle with a thickness a lot less than that attributed to a tectonic plate.

Alternative explanation for the formation of mid-ocean ridge systems

The thickness of the Earth's crust varies between 20 and 90 km beneath continents with an average of 35 km. Mountains on land often rise to 6 – 7 km above sea level. The crust of ocean basins is on average only 7 km thick. Thick continental crust weighs heavily upon the mantle and indeed is implanted deep into the mantle. I propose that this causes the Earth's mantle beneath continents to be compressed; to compensate for this compression over large areas, the mantle beneath the ocean basins – where the crust is much thinner – decompresses. The decompression of the mantle down the middle of oceanic basins has led to seams of magma chambers that erupt through fault lines circling the planet. These fissures of erupting magma have built up extensive mid-ocean ridge systems.

I propose that the decompression of the upper mantle below oceanic crust has been aided by the formation of transform fault lines at right angles to the ridges. This transverse faulting has divided the ridges into sections or slabs that can detach themselves from the surrounding crust. The extensive transverse faulting has had the effect of 'lifting the lid' of the thin crust in the middle of oceanic basins. The 'lifting of the lid' has allowed the mantle to decompress, expand upwards and form molten magma that erupts constantly from the fissures of mid-oceanic ridge systems.

See diagram 21.3 in the section on Mid-Ocean ridges.

Magnetic reversals

One of the main pieces of evidence used in support of Sea Floor Spreading is data from palaeomagnetism. The Earth's magnetic field has been reversed from time to time such that magnetic north did not point in the direction that it points now. The direction of the Earth's magnetic field is recorded in basalt rock since as it solidifies at the surface, the iron minerals it contains align with the Earth's magnetic field and remain set that way.

According to the Sea Floor Spreading Model reversals in the Earth's magnetic field recorded in sea floor basalts show that new ocean floor is being produced at the mid-Atlantic ridge at its fastest speed at 3 cm per year (Zeilik 2002, page 158-159).

Let us examine both the observation and theory behind magnetic reversals more closely. On land magnetic reversals in rocks show a confusing blobby pattern on the map. In contrast, the sea floor of oceans shows a pattern consisting of a set of magnetic bands alternately higher and lower than the average magnetic field today. These positive and negative anomalies lie in a linear pattern parallel to mid-ocean ridge axes, although they are offset along the fracture zones of faults.

In a zone maybe over 500 km wide either side of the axis of a mid-ocean ridge system there are elevated ridges and isolated peaks that have what is called normal polarity. These ridges and peaks are interspersed by dips with some sediment infilling that have reversed polarity. Thus, a stripy pattern in magnetic reversals is observed. The pattern on one side is the approximate mirror image of that on the other side of the axis.

In 1963 Drummond Matthews and Frederick Vine [21] postulated that new oceanic crust was being formed along the axis of the mid-ocean ridge and taking on the polarity of the Earth's magnetic field at the time of its formation. Each new block would split and the two halves move to each side generating the bilateral magnetic symmetry. If crust is being produced at a constant rate, the widths of individual magnetic polarities would correspond to the intervals of time between magnetic reversals. The Vine-Matthews Hypothesis was correlated

---

[21] Vine, Frederick J. & Drummond H. Matthews (1963) Magnetic Anomalies Over Oceanic Ridges *Nature* Vol. 199 (4897), pages 947-949.

with a time scale for magnetic reversals published by some American geologists including Allan Cox in 1966. Allan Cox stated that reversals of Earth's magnetic field in the structure of the sea floor can be read like tree rings.

The magnetic anomaly time scale has now been extended back to the Cretaceous period. It is thought that about every million years, magnetic north flips to point south. Scientists have used the pattern of magnetic reversals to work out the age of ocean floor. Magnetic anomalies have been used to show that none of the rocks of the deep ocean floor are more than 200 million years old.

The sea floor either side of a mid-ocean ridge is alternately positively and negatively magnetized. The Sea Floor Spreading Model presents this as ridges being produced in a chronological sequence. According to this model, each time the Earth's magnetic field reverses a new stripe is added on each side. However, if this is the case, why do the ridges and peaks have normal polarity and the troughs reversed polarity? Surely a chronological sequence should involve all levels of surface, not a pattern that conforms to the height of the sea floor?

I suggest that all the ridges of a mid-ocean ridge system have been produced simultaneously and that is why they all have the same magnetic polarity. Magma appears to have erupted from rifts along the axes of ridges, each longitudinal vent of magma flow building up its own ridge. The ridges may follow many parallel fault lines all with the same orientation in a given part of the ocean. The lower lying sea floor with reversed polarity in between the ridges would have been produced at an earlier time simultaneously when the Earth's magnetic field pointed the other way.

I am suggesting that the elevated ridges belonging to mid-ocean ridge systems that form wide zones (sometimes 1000 km wide) are of similar age. This idea would make them more similar to mountain ranges found on land. A particular mountain range consisting of many peaks is usually attributed as having been formed in a particular geological period simultaneously. The peaks of mountain ranges on land are not thought to have formed consecutively. Mountain ranges on land are often associated with faulting. The fault lines may have run parallel to each other in a particular orientation as they do on the sea floor. However, mountain ranges on continents have been subject to additional forces such as the folding of sedimentary strata, and to extensive weathering and erosion. The end result of this would produce a more uneven pattern.

If the high ground of the sea floor dates from a more recent period of geological history (as do the present highest mountain ranges on land) and the lower ground dates from a former period, then the Earth's magnetic field has not reversed every million years. There may have been a few changes to the Earth's magnetic field over Earth history, not many reversals.

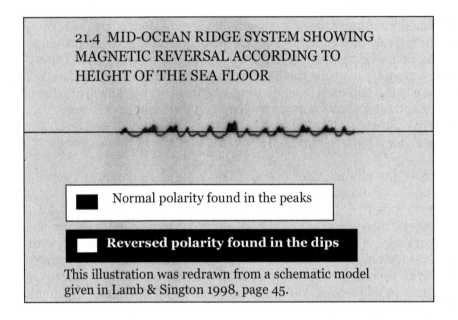

21.4 MID-OCEAN RIDGE SYSTEM SHOWING MAGNETIC REVERSAL ACCORDING TO HEIGHT OF THE SEA FLOOR

■ Normal polarity found in the peaks

□ **Reversed polarity found in the dips**

This illustration was redrawn from a schematic model given in Lamb & Sington 1998, page 45.

Deep sea sediments

The age of deep sea sediments that contain fossil microorganisms have been found to be older away from the mid-ocean ridges. This has been taken as confirmation of the Sea Floor Spreading Hypothesis.

Where continents meet the oceans at continental shelves there are thick sediments washed from the land by rivers. Continental shelves are inclined abruptly where they meet the deep ocean that has water 4 km deep. The sea floor of the deep ocean is known as the smooth abyssal plains. It is smooth because layers of sediment cover the volcanic crust. The zone of mid-ocean ridges have some sediments filling in the troughs, but little sediment overall.

The axis of the mid-ocean ridges of the South Atlantic Ocean has sediments of Pleistocene age around the central ridge, and sediments of Pliocene, Miocene, Oligocene and Eocene further out (Lamb & Sington 1998, page 54).

I do not think that the distribution of sediments of different age proves the Sea Floor Spreading Model because if the basalt of the mid-ocean ridges is fairly recently erupted, it could not be covered with older sediments. The freshly deposited sediments could only have been deposited after the ridges were formed. Also, the raising up of the sea floor by volcanic activity is likely to expose any older sediment to being eroded away by ocean currents. That layers of sediment do exist in the abyssal plains where volcanic activity is not taking place now is what one would expect –sediment remains where it has not been disturbed.

The hypothesis that the ridges beneath the oceans have been produced as a chronological sequence due to sea floor spreading has meant that much of the data relating to the ocean floors is theory-dependent data. This theory-dependent data is used to date the ocean floor which confirms the original theory. One must be aware of how data is generated. I do not believe that the oceans have been formed by widening gaps between parts of a supercontinent. I do, however, believe that the basalt rocks of the ocean floors are comparatively recent, but for different reasons.

Age of the ocean floor
According to the Sea Floor Spreading Model, the Atlantic Ocean floor is 125 million years old. This would mean that the continents of South America and Africa were touching during the Cretaceous period, and have since rifted apart. No ocean floor is thought to be older than 200 million years.

We have noted elsewhere that oceanic crust is made of basalt on average 7 km thick covered by varying depths of sediments. Where basalt emerges at mid-ocean ridges, the ocean floor is much higher than in the vicinity of oceanic trenches; the ocean floor becomes progressively deeper away from the central crest of mid-ocean ridges.

Ocean floors are young in comparison with the sedimentary strata of continents.

If it is accepted that oceanic crust is much younger than continental crust formed mostly from sedimentary strata and granite masses – why is this so?

The evidence shows that landmasses now exist where shallow seas once existed. The ancient sedimentary strata laid down in seas of the Precambrian, Palaeozoic and Mesozoic eras have been uplifted by folding and faulting to form the present day landmasses.

The present ocean basins have oceanic crust formed from basalt that continues to erupt from ridges along parallel fault lines. I contest that it is the constant weathering of basalt that means that it will never accumulate to very great thicknesses. The weathering of basalt is probably mainly mediated by microbes (as explained in Chapter 18, section 7) under aerobic conditions. The eating away of basalt by microbes means that basalt cannot be old in geological terms.

Flood basalts on land are not older than 250 million years – the Siberian Traps date from 250 million years, and all other plateaus formed from basalt on land are younger (see Chapter 19, section 5). Older basalts on land such as in Greenstone Belts have been preserved as metamorphic rocks by heat from volcanic sources - metamorphic rocks are harder and are not weathered like other sedimentary rocks. Thus, like the ocean floor, there are no very old basalts on land.

The sediments accumulating on the abyssal plains of the deep ocean floor date mostly from the present Cenozoic era. The sediments that accumulate on the ocean floor will eventually become consolidated into sedimentary strata. The most ancient sedimentary strata formed as ocean floor have been uplifted to form land.

6. Sea level

Over geological time the depth of the oceans has increased. The average depth of oceans today is 4000 metres, although above mid-ocean ridges the depth of water may only be between 1500 and 3000 metres, while at oceanic trenches it may be between 6000 and 8000 metres. The seas of present-day continental shelves are 200 metres deep or less.

The oceans of the Earth were once shallower than they are now. Deep oceans of 4000 metres are a geologically young feature of the planet. Evidence that the oceans were once shallow is that fossil coral has been dredged up from the seabed of the Pacific Ocean from depths of several thousands of metres. These coral fragments were once part of a coral reef that lived 100 million years ago in shallow water (Lamb & Sington 1998, page 41).

Sea level has risen by 10-25 cm during the twentieth century. This is thought to be due to global warming and the melting of glaciers. The recent increase in sea level is a separate phenomenon to the rise in sea level in terms of Earth history.

The long-term increase in the depth of the oceans has been attributed to the notion that ocean floors are sinking. It is generally believed today that the ocean basins are basins due to subsidence. The idea that the ocean floors are sinking is part of the Sea Floor Spreading Model.

Ocean basin subsidence and coral reefs
Charles Darwin may have been the first person to suggest that the ocean floor is sinking. During the voyage of the Beagle Darwin maintained his interest in geology as well as in the distribution of species. It was when the Beagle was heading for home in 1836 crossing the Indian Ocean that Darwin had the opportunity of making a study of coral islands. This led him to think up a general theory involving oceanic subsidence. Darwin collected information about all that was known on the structure and distribution of coral reefs. He reached the conclusion that over vast stretches of the Pacific and Indian Oceans an ancient land surface had slowly sunk beneath the water, and that the mountain peaks of this former land surface were to be recognized in the countless groups and archipelagos of coral islands

in these oceans (Geikie 1909 reprint 2009, page 32-34). Darwin made special reference to coral atolls in his theory.

An atoll is a ring of coral reefs that rises from the abyssal floor of the ocean and surrounds a lagoon that may be 50 metres deep or more. Corals live at a depth of down to 30 metres, but the reefs forming atolls go down much deeper to well over 300 metres deep. The question is, how can corals that only live to 30 metres depth, form a reef with over 300 metres depth?

Darwin's hypothesis was that as the seabed subsided, fringing reefs of coral along a coastline would be converted into barrier-reefs with a lagoon-channel between the reef and the shore. In the case of islands, the island itself would continue to sink with the surrounding ocean floor while at the same time the coral polyps in their accumulation of calcareous material would keep apace with the downward movement by growing upwards. The result would be that the reef would become an atoll or ring of coral rock enclosing a lagoon beneath which the last peak of land might eventually disappear.

Darwin presented his Theory of Coral Reefs before the Geological Society in 1837 and first published it in 1842 [22].

Darwin's views attained general acceptance. However, the first objection to the general applicability of Darwin's explanation of coral islands appears to have been raised by Louis Agassiz in 1851 after Agassiz' study of coral reefs in Florida.

It was found that in parts of the Pacific and Indian Oceans as well as in the warmer waters of the Western Atlantic Ocean there is evidence of uplift of the ocean floor where many groups of coral islands are found; the elevation of the ocean floor amounting in some cases to more than 1000 feet (304.8 metres) (Geikie 1909 reprint 2009, page 35). Archibald Geikie and other geologists at the beginning of the 20th century believed that there was evidence for elevation of the ocean floor, where according to Darwin's theory there should be depression.

---

[22] Charles Darwin (1842) *First Part of the Geology of the Beagle: The Structure and Distribution of Coral Reefs*. A newer edition of the theory came out in 1874 entitled *Coral Reefs*.

These opponents also noted that coral islands are almost all of volcanic origin, and volcanic activity is associated with a raising of the level of the ocean floor.

The alternative theory to Darwin's theory involved the idea that coral islands emerged from the sea by elevation of the ocean floor. If this were true the formation of barrier-reefs could only be explained by the idea that bits of the reef kept breaking off the top of the reef and accumulating lower down on the seabed. This would form a reef of broken pieces.

In order to settle the issue, borings were carried out on the atoll Funafuti by the Royal Society in 1904. The borings revealed reef-building coral polyps that were intact from the top to the bottom of the reef over 1100 feet (over 330 metres) down. The evidence showed a mass of coral rock that appeared to be found in its original position of growth, with no signs of it being formed of broken pieces (Geikie 1909 reprint 2009, pages 81-82). Therefore, at the turn of the century Darwin's hypothesis that coral islands are evidence of subsidence in ocean basins won the day.

Rising sea level and coral reefs

Darwin was, of course, right when he observed that coral polyps grow upwards always maintaining the top of the reef at the same depth where sunlight penetrates just below water level. His deduction that coral reefs hundreds of metres deep must be formed by a constant accumulation from upward growth is unavoidable. Darwin's explanation was subsidence of the ocean floor with the sea level remaining constant – in this model sea level cannot go down with the deepening of the basin.

A logical alternative explanation is that sea level has risen because the amount of water contained in the oceans has increased. If sea level rose slowly over geological time with shallow seas becoming deep oceans, the continuing growth of coral polyps would form reefs as we observe them today.

An atoll could be formed by the sea level rising and the reef growing upwards to maintain itself just below water level. This would have the consequence of increasing the height of the reef over time. An ancient volcanic island in the middle of the lagoon would become smaller as it became inundated by the rising water. Eventually the tip of the island would disappear beneath sea level and only the lagoon would be left. At this stage the island becomes an atoll.

Guyots

It is generally believed today that tracts of ocean basins are subsiding although it is not believed, as Darwin did, that they represent subsiding ancient land surfaces. The theme of ocean floors sinking was taken up by Harry Hess when he proposed that features that he named 'guyots' are proof of the Sea Floor Spreading Model.

Another name for guyots is seamounts. Seamounts are extinct submarine volcanoes that are conically shaped and often flat-topped. They rise abruptly from the abyssal plain to heights of 1100 metres or more above the ocean floor, but they do not reach the surface of the ocean. The Pacific and Indian Oceans are strewn with thousands of these underwater mountains.

Flat-topped seamounts were given the name guyots by Hess in 1960. Hess recognized that guyots resemble low Pacific islands; Pacific islands are volcanic cones that are flat on top because their tops have been removed by wave erosion at the sea surface. The guyots beneath the sea always reach to about the same height above the surrounding ocean floor, but their flat tops are thousands of metres below sea level. It would appear that guyots were once volcanic islands whose tops were eroded flat by wave erosion at sea level, but which now find themselves covered by water.

Hess conceived the idea that guyots were formed as volcanic islands in the vicinity of mid-ocean ridges where they were eroded flat at the sea surface. These islands then moved across the ocean with the lateral spreading of the ocean floor and were carried down below sea level by the sinking of the ocean floor. He proposed that as the basalt ocean floor cools down away from mid-ocean ridges, the cooling causes it to contract and become denser, and thus sink down. In his model guyots formed in the middle of the ocean would move 1500 km laterally in 150 million years, moving into deeper water towards the oceanic trenches. Thus, guyots moving on an oceanic conveyor belt are taken as proof of the Sea Floor Spreading Model.

Comment
The presence of guyots below the surface of the water cannot be explained, in my opinion, simply by subsidence of the ocean floor since if the ocean floor sank, but the amount of water remained the same, the water level would simply go down in comparison with land; the ocean would not become deeper. In this case, if guyots went down with a sinking ocean floor, the level of water would also go down, and the guyots would remain as islands at the surface of the water.

Of course, what Hess proposed was that volcanic islands migrate across ocean basins moving into deeper water. I do not believe that islands migrate across the globe.

In fact, the presence of guyots or flat-topped seamounts beneath the surface of oceans is proof of two hypotheses that I am proposing. It is proof firstly that the amount of water on Earth has increased, and is probably still increasing (the thesis of Chapter 10 of Part I and section 8 of Chapter 18 of Part III). Secondly, it is evidence that our planet has shrunk in size (the thesis of the present Chapter 21 of Part III).

Reduction in ocean basins and water sources
In accordance with the Earth Shrinkage Hypothesis, I propose that one of the explanations for the generalized rise in sea level over geological time is that the size of ocean basins has shrunk.

Deep sea trenches are zones of subduction where oceanic crust disappears into the mantle. Trenches act as tucks in the Earth's surface that affect the oceanic crust, but not continental crust. As sea floor disappears, the oceans must become smaller, and indeed the whole planet must become smaller. As ocean basins become smaller, they will have a higher sea level.

The sources of water on Earth were discussed in Chapter 18, section 8. I concluded that water on Earth has been and still is being produced by subsurface bacterial communities living beneath the ocean floor, and by bacteria living on basalt minerals belonging to the ocean floor. Combustion of organic matter contained in marine sediments in connection with volcanism is also very significant. The amount of water emerging from hydrothermal vents in the vicinity of mid-ocean ridges is significant.

The thread running through various chapters in this book is the idea that the Earth started off with very little water and resembled other terrestrial planets in the solar system which have surfaces that are generally dry. The Earth has acquired water through the processes of life. The increase in the amount of water on Earth has raised sea level in the oceans over time.

The thesis of this section is that the Earth was once larger than it is now, and being larger it had shallower oceans.

Conclusion

The Earth Shrinkage Hypothesis shows how as the Earth's core cools down and shrinks in size, the unshrinkableness of the Earth's crust has led to a surface that is wrinkled, rent and tucked to produce mountain ranges as well as mid-oceanic ridge systems and oceanic trenches. The large numbers of earthquakes occurring each year are testimony to the forces involved.

Former ocean floors such as the Tethys Ocean floor have been converted into mountain ranges. Continental landmasses formed of sedimentary strata and granite intrusions have been pushed up, while ocean basins of basalt have undergone various processes of reduction.

The Earth's mantle has adjusted to the shrinking core by decompression forming magma chambers in the upper mantle. These magma chambers evacuate molten magma onto the Earth's surface in volcanic eruptions and outflows along fault lines forming basalt ridges on ocean floors. In this way volcanism moves material from the mantle to the surface adjusting the size of the Earth's mantle.

The Earth Shrinkage Hypothesis is offered as an explanation for geological processes occurring on the face of the Earth as an alternative to Plate Tectonics Theory.

In the model of the Earth provided by the Earth Shrinkage Hypothesis the mantle is not divided into lithosphere and asthenosphere. The mantle is increasingly hotter with depth, but the mantle is solid due to compression except where liquid magma chambers have formed in the upper mantle. This is shown by seismological studies – the analysis of the vibrational waves produced by earthquakes.

Under the Earth Shrinkage Hypothesis geological processes including subduction are viewed as involving the Earth's crust only i.e. much thinner slabs of rock than the 50-100 km thick lithosphere plates invoked by Plate Tectonics. Under the Earth Shrinkage Hypothesis there is no distinct lithosphere divided into plates moving upon an asthenosphere molten mantle.

Concerning sea level, Darwin's explanation for the formation of coral islands and atolls lends strong support to my own hypothesis that sea level has risen over geological time due to an increase in the amount of water on Earth and a reduction in the size of ocean basins; the explanation is not a proof of oceanic subsidence.

The existence of guyots or flat-topped seamounts that were once islands, but are now found thousands of metres below sea level is convincing evidence that the oceans were once fairly shallow – about 1000 metres deep on average compared to their present depth of 4000 metres for the abyssal plains. This does not prove that guyots have been migrating across ocean basins moving into deeper water; it proves that sea level has risen.

News reports constantly come to us of disaster – events affecting the Earth's crust – volcanic eruptions, earth quakes and tsunamis. We look, but only theory allows us to see what is happening. Theory is understanding.

The human cost of natural disasters can only be avoided by warning systems and better constructed buildings. Divine retribution was a notion that was always associated with the occurrence of natural disasters (when it happened to others, not to ourselves). Today the prayers and efforts of religious people are directed towards relief and aid for victims of disasters.

Bibliography

Encyclopedia Britannica 2011 Standard Edition: Atoll; Continental shelf; Deep-sea trench; Earth; Earthquake; Himalayas; Indian Ocean; Island arc; Metallurgy; Ocean; Plate tectonics; San Andreas Fault; Seismic wave; Tethys; Volcanism; Volcano

Geikie, Archibald (2009) *Charles Darwin as Geologist: The Rede Lecture, given at the Darwin Centennial Commemoration on 24 June 1909* Cambridge University Press

Ince, Martin (2007) *The Rough Guide to the Earth* Rough Guides Penguin Group

Lamb, Simon & David Sington (1998) *Earth Story: The Shaping of Our World* BBC Books

Taylor, Stuart Ross (1998) *Destiny or Chance: Our Solar System and its Place in the Cosmos* Cambridge University Press

Zeilik, Michael (2002) *Astronomy: The Evolving Universe* 9th edition Cambridge University Press

CHAPTER 22

GEODYNAMOS

Introduction

This last chapter is about geodynamos and magnetic fields. The Earth's mechanism for production of a magnetic field is vital to life on Earth and is closely related to Earth's internal structure and its mode of formation. The Sun's magnetic field which envelopes the whole of the solar system interacts with Earth's magnetic field to produce the magnetosphere.

In looking at dynamo mechanisms for producing magnetic fields, further evidence will be examined in support of the hypotheses presented in Part III.

1. Magnetic fields

Stars such as the Sun, gaseous planets such as Jupiter and terrestrial planets such as Earth have magnetic fields. Earth's magnetic field is much stronger than that of other terrestrial planets in the solar system.

This first section describes the magnetic fields of the Sun and Earth.

Sunspots

The outer layers of the Sun consist of the photosphere, surrounded by a thin chromosphere and this is surrounded by the corona which extends for millions of kilometres into space. Sunspots are cool regions of 3800 K embedded within the photosphere which has a general temperature of 5800 K.

The Sun's magnetic field emerges from sunspots. Sunspots produce the Sun's corona which is a halo of particles around the Sun extending beyond the planets and flowing out at a speed of 400 kilometres per second near the Earth. The charged particles of this flow are known as the **solar wind**. The particles of the corona or solar wind may reach a temperature of a million K.

The solar wind drags magnetic field lines out from the surface of the Sun. Travelling at an average speed of 500 km/second the particles reach the orbit of Saturn in one solar rotation of 27 days. This means that a source of particles on the Sun has gone completely round in this time period and for this reason the magnetic field lines emanating from the Sun describe a spiral. The magnetic field lines do not break and particles move along these paths. It takes four days for the solar wind to reach the Earth having originated from a point that has rotated 50 degrees from its original position facing the Earth.

There is a constant ebb and flow of sunspot activity on the Sun which is thought to influence Earth's climate.

Sunspots are regions of extremely strong magnetic field which develop into the shape of a daisy. The dark central core is cool having a temperature of 3000 K. A magnetic flux loop emerges from the core and spreads outward. The smallest size of lasting sunspots is about 500 kilometres diameter, while the largest sunspots can be 50 000 km in diameter. Sunspots form in pairs of opposite polarity often as groups of opposite polarity. Clusters of magnetic flux loops emerge from below the surface in pairs of opposite polarity connected by dark arches in the chronosphere above.

The sunspot pairs in one hemisphere of the Sun all have the spot towards the direction of rotation –the leading spot with one polarity and the following spot with the opposite polarity. Spots that develop with the opposite polarity configuration for the hemisphere usually die out, but occasionally regions of reversed polarity grow into large, active spot groups.

The average lifetime of an individual spot group is about one solar rotation (27 days). But the number of sunspots increases over an 11 year period. At sunspot maximum there may be 10 groups and 300 spots across the Sun, then the sunspot cycle returns to minimum. For a given 11-year cycle, the magnetic polarity configuration of spot groups is the same in a given hemisphere and is reversed in the opposite hemisphere. In the next cycle the magnetic polarity configuration of each hemisphere reverses.

Since the magnetic polarity configuration of sunspot groups reverses every 11 years, it returns to the same value every 22 years. This is the period of the complete magnetic cycle.

Sunspots on the equator of the Sun rotate around at a 25-day rate, while sunspots at high latitudes rotate at a 28 or 29-day rate. Thus, the surface of the Sun rotates faster at the equator than at the poles, while the interior below the convective zone rotates as a solid body. This differential rotation is thought to play an important role in the generation of the magnetic field. The relative motions in the Sun may twist and enhance magnetic flux loops.

A **solar flare** is a sudden release of magnetic energy from a sunspot region. The energy released in the three or four biggest flares each year is equivalent to all the energy produced in small flares over the year. A flare can be likened to a giant natural synchrotron accelerating vast numbers of electrons to energies above 10 000 electron volts (10 KeV) and protons to above one million electron volts (1 MeV) (*Encyclopedia Britannica 2011 Standard Edition*: Sun: Flares). The energy input into electrons produces X-ray bursts and radio bursts, and heats the Sun's atmosphere; and the energy input into protons produces gamma-ray lines followed by X-rays. Solar flares are hardly seen in visible light.

Solar flares occur along neutral lines which form boundaries between regions of opposite magnetic polarity. The distortion of the field when relative motion or rapid flux emergence occurs produces energy which is suddenly released in flares. Impulsive flares produce outward explosion and ejection of material – great clouds of coronal material are blown out and make up a large part of the solar wind.

The low-density plasma of the corona or solar wind radiates very little and this allows it to reach and maintain high temperatures. The temperature of the interplanetary medium is over 200 000 K near the Earth.

Earth's magnetosphere
Besides having a rotational axis, the Earth also has a magnetic axis. The magnetic axis connects the magnetic north pole with the magnetic south pole and is indicated by the needle of a compass. The magnetic

axis emanating from the core produces a magnetic field around the Earth.

Where the Sun's magnetic field meets the Earth's magnetic field, the Earth's magnetosphere is formed. The magnetosphere consists of a plasma of protons and electrons from the solar wind or Sun's corona trapped by the Earth's magnetic field. It prevents solar cosmic rays carried in the solar wind from reaching the Earth's surface where they would cause damage to life.

The magnetosphere is teardrop-shaped and goes out to a distance of 65 000 km on the Sun side where the pressure of the solar wind is balanced by the geomagnetic field. This serves as an obstacle to the solar wind and the flow of charged particles is deflected around the Earth by the resulting bow shock. The solar wind plasma streams out into an elongated magnetotail stretching for several million kilometres downstream from Earth away from the Sun.

Close to the Earth two doughnut-shaped belts of trapped protons and electrons are formed called the Van Allen radiation belts.

When the Sun's interplanetary magnetic field switches to a direction opposite to the Earth's field, or when big clouds of particles released by solar flares enter the Earth's magnetosphere, the fields in the magnetotail reconnect and release energy. This produces the aurora borealis or northern lights. Large solar flares which eject large quantities of energetic particles into the solar wind may form a ring current around the magnetosphere. This produces sharp fluctuations known as geomagnetic storms in the Earth's field. Geomagnetic storms disturb radio communication and produce voltage surges in long-distance transmission lines.

Reversals in Earth's magnetic field
The Earth's magnetic field resembles the magnetic field produced by a giant permanent magnet or bar magnet at the centre of the Earth. At the present time the north magnetic pole is located at the south geographic pole of the Earth. This is termed 'normal polarity'. Normal polarity has not always been the case on Earth; many times in the history of the Earth the direction of the magnetic field has pointed the opposite way. When magnetic north corresponds to geographic north it is termed reversed polarity. (Use of the terms 'normal

228

polarity' and 'reversed polarity' is slightly confusing when it is up-side-down to what one would assume to be the case). A compass needle is 'north seeking' which means that it apparently points more or less to geographic north, when in fact, it points to magnetic north which is geographic south.

The imprint of the Earth's magnetic field is preserved in rocks containing iron-rich minerals such as magnetite and haematite. Volcanic rocks such as basalt are good recorders of palaeomagnetism, and some sediments also align their ferromagnetic particles with the Earth's magnetic field at the time of deposition. Both rock strata and deep-sea cores have shown reversals in magnetic north. Radiometric dating methods such as potassium-argon have been used to add a geologic time scale to reversals.

Evidence for magnetic reversals has been taken from surveys made by ship taking samples of ocean floor from the Mid-Atlantic Ridge system. The samples show a stripy pattern of alternate normal and reversed polarity. The Vine-Matthews Hypothesis (explained in Chapter 21, section 5 on magnetic reversals) proposes that the stripes were produced chronologically on either side of the mid-oceanic ridge and therefore can be read like a magnetic tape capturing the alternating sequence of the Earth's magnetic field orientations. This leads to the conclusion that magnetic reversals have occurred every 300 000 to one million years during the present Cenozoic era.

I have proposed in Chapter 21 with respect to the Vine-Matthews Hypothesis that since the higher level ridges of the Mid-Atlantic Ridge system all have normal polarity and the lower level dips all have reversed polarity, that all the ridges were formed during the *same* time period and represent recently-formed ocean floor, while the dips represent underlying older ocean floor. If this were true, magnetic reversals would not have occurred every million years or so, but would have occurred over much more intermittent time periods.

No regularities or periodicities have been found in the pattern of field reversals. The chart for geomagnetic reversal timescales shows that during the present Cenozoic era and beyond there have been 184 polarity intervals in 83 million years; during the Mesozoic era between 83 and about 120 million years ago there was no reversal for 40 million years – this is termed Cretaceous Normal; during the Palaeozoic era between 262 and 312 million years ago there was no reversal for over 50 million years – this is called the Kiaman Reverse Superchron of the Late Carboniferous to Late Permian periods. According to the interpretation I have proposed, the Cenozoic era would not have had 184 polarity intervals; it may have had two or not many more which would make it more similar to other eras.

Polar wandering

Rocks that underwent magnetization as they cooled provide fossil compasses that reveal the direction of the magnetic pole and the latitude of the rock strata sample as it cooled. Rocks formed at the magnetic equator have horizontal magnetization, while rocks formed at higher magnetic latitudes contain a field pointing up or down at an inclination that depends on latitude.

Using the angle of dipole in rock samples it is possible to infer the location of a virtual magnetic pole relative to the location of the sample. Samples of rocks dated to younger than 20 million years do not depart from present magnetic pole locations by distances greater than experimental uncertainties. For samples older than 30 million years, however, successively greater 'virtual pole' distances are revealed. This suggests that the magnetic poles have wandered over time.

Polar wandering was attributed as a characteristic of the Earth's magnetic field until recently. However, due to discrepancies in the polar wandering curves for different continents it was suggested that the data shows that the continents have wandered over the globe. This would mean that the continents have changed their latitude, while the magnetic poles have not moved significantly relative to the geographic poles. The new interpretation of the data supports the Theory of Continental Drift rather than the notion that the magnetic poles have changed location. Pole location curves from younger samples converge to the present pole location.

As I stated in Chapter 21, I do not accept the Theory of Continental Drift. However, I believe that the Earth has shrunk in size and that oceanic crust has disappeared down oceanic trenches. Therefore, the surface of the Earth has not remained fixed and constant in terms of GPS locations. The picture is likely to be a complex one.

Wavering in Earth's magnetic field

The Earth's magnetic field shows short-term variations both in direction and strength. Scientific observations of the magnetic field do not go far back, although the magnetic minerals in articles of fired clay provide some additional clues.

Observations of the declination (angle) of magnetic north have been made in London since 1540. These observations show that the direction of the field at that site has varied over 30 ° (*Encyclopedia Britannica 2011 Standard Edition*: Geomagnetic field: Secular variation of the main field). The strength of the dipole magnetic field has been shown to be decreasing since 1850. If this trend continues the magnetic field emanating from the Earth's core may vanish in another 2000 years.

The orientation of the Earth's dipole appears to be drifting westward at a rate of 0.08 ° per year. The Earth's magnetic field, apart from the dipole coming from the core, has components which emanate from the atmosphere and surrounding magnetosphere. Some non-dipole components of the field appear to be drifting westward at an average rate of 0.18 ° per year. At this rate the drifting features would circle the Earth in 2000 years.

Measurements of the Earth's magnetic field suggest that the Earth's dipole may be in the process of reversing. The absence of a dipole component would mean that the solar wind would approach the surface of the Earth and cosmic ray particles which it contains would hit the Earth instead of being deflected away by the Earth's magnetic field or trapped in the magnetosphere. Decaying of the magnetic field would allow cosmic ray particles to cause genetic damage to plants, animals and humans.

2. Ferromagnetism and induced fields

Magnetic fields are produced by ferromagnetic materials which include iron, nickel and cobalt, or by electromagnetic induction involving electric currents. During the 19th century famous discoveries were made concerning magnetism and electricity.

The Curie point

Magnetism is a property of solid iron, nickel and cobalt. Magnetism in these ferromagnetic materials appears when they cool down below the Curie temperature or Curie point. For pure iron the Curie point is 770 °C and for rocks containing the mineral magnetite it is 570 °C. One of the highest Curie points is for cobalt at 1121 °C.

The Curie point is named after the French physicist Pierre Curie (1859-1906). Curie defended his thesis on magnetism and obtained a doctorate of science in 1895, the same year in which he married Polish-born Marie. Pierre and Marie Curie along with Henri Becquerel were awarded the Nobel Prize for Physics in 1903.

The atoms of ferromagnetic materials behave as tiny magnets. Below 912 °C iron forms a body-centred crystal structure known as alpha-ferrite (see appendix A for Chapter 21). Below 770 °C the iron atoms spontaneously align themselves; they all become orientated within each domain in the same direction such that their magnetic fields reinforce each other.

In antiferromagnetic materials, atomic magnets alternate in opposite directions such that their magnetic fields cancel each other out.

Ferrimagnetic materials usually contain two different types of magnetic atoms; the spontaneous arrangement is a combination of both patterns such that there is only a partial reinforcement of magnetic fields.

If the temperature is raised to the Curie point for ferromagnetic, antiferromagnetic or ferrimagnetic materials the spontaneous arrangement of atoms is disrupted and only a weak kind of magnetic behaviour remains called paramagnetism. Increasing the temperature above the Curie point produces decreasing paramagnetism. When these materials are cooled down again below the Curie point, the magnetic atoms spontaneously realign themselves and ferromagnetism, antiferromagnetism or ferrimagnetism reappears.

Electromagnetic induction

Magnetic fields can also be produced by induction with electric currents. An electric field is produced by electric charges at rest relative to an observer. A magnetic field is produced by moving electric charges.

The sense of the magnetic field depends on the direction of the electric current which is defined as the direction of motion of positive charges. The right-hand rule defines the direction of the magnetic field by stating that it points in the direction of the fingers of the right hand when the thumb points in the direction of the current.

Michael Faraday

Electromagnetic induction was discovered by the English physicist and chemist Michael Faraday (1791-1867) during the 1830s. It became the principle upon which energy-conversion devices such as electric generators are based.

While working on magnetism, Faraday found that moving a permanent magnet into and out of a coil of wire induced an electric current in the wire. This was developed into a means of converting mechanical energy into electricity on a large scale.

The story of Faraday is worth the telling: Michael Faraday came from humble beginnings as the son of a blacksmith in Surrey. He received the rudiments of an education in a church Sunday school and started working doing deliveries for a bookbinder at a young age. When he was 14 he became apprenticed to the bookbinder. The young Faraday started to read the books on science that had been sent for rebinding. He became self-taught to a level that enabled him to find work as a laboratory assistant to Sir Humphry Davy in 1812. The laboratory gave him the opportunity to start conducting his own experiments. Faraday became one of the greatest scientists of the 19th century making ground-breaking discoveries in the understanding of electricity and chemistry. Faraday's life was guided by his adherence to a Christian sect called the Sandemanians who were an off-shoot of the Scottish Presbyterians. Surely this story should disarm those who see an opposition between the holding of strong religious beliefs and science.

Types of field
A magnetic field may be a primary field, an induced field or a
remanent field.

The Earth's magnetic field is a primary field. It is not known how it is
produced. This is the subject of the sections that follow in this
chapter. The Earth's magnetic field resembles the dipolar field
produced by a giant iron permanent magnet at the centre of the Earth
with the south pole of this magnet oscillating around the North Pole.

An induced magnetic field has magnetization in proportion to the
strength of the inducing field. An induced field vanishes when the
primary field vanishes.

The magnetic field of the Earth's atmosphere and magnetosphere
would vanish if the Earth's primary field vanished. Apart from the
dipole emanating from the core, the Earth's magnetic field is produced
by the ionospheric dynamo, the ring current, the magnetopause
current, the tail current, field-aligned currents and auroral electrojets.

At the Earth's surface 90 % of the magnetic field arises from sources
internal to the planet, whereas above the Earth's surface the effect of
other sources becomes as strong as the field from the Earth's core.
However, the non-dipole sources of Earth's magnetic field would not
exist without the geomagnetic dynamo in the Earth's interior.

A remanent magnetic field is produced in ferromagnetic materials by a
primary field, and remains in these materials after the primary field
has disappeared. Remanent magnetization is created by
ferromagnetic atoms forming magnetic domains – regions of aligned
dipoles held in place by interatomic forces.

The basalt rocks of the Earth's crust have remanent magnetism
acquired by trapping the dipole alignment of the Earth's main field
when the molten magma cooled and hardened into rock. Crustal
magnetization is responsible for some of Earth's magnetic field.

3. Dynamo Theory

William Gilbert (1544-1603) physician to Queen Elizabeth I and to King James I in England concluded that the Earth is magnetic. He published *De Magnete, Magneticisque Corporibus, et de Magno Magnete Tellure* (On the Magnet, Magnetic Bodies, and the Great Magnet of the Earth) in 1600. This work gives a full account of his research on magnetic bodies and electrical attractions. Gilbert proposed that a compass needle points north-south and dips downward because the Earth acts as a lodestone (which is a magnetic oxide of iron) or bar magnet.

The internal origin of the Earth's magnetic field was demonstrated by the German mathematician and astronomer Carl Friedrich Gauss (1777-1855) in the 1830s.

During the 20th century it was proposed by various people including Joseph Larmor, Walter M. Elsasser and Sir Edward Bullard that the Earth's magnetic field did not result from the permanent magnetism of ferromagnetic materials, but was generated by a dynamo mechanism in the Earth's fluid outer core.

The reason for discounting the idea that the Earth's magnetic field arises from a permanent magnet is that the Curie point at which ferromagnetic materials lose their magnetic properties is reached only about 20 kilometres beneath the Earth's surface. This means that although the Earth's crust has remanent magnetism, most of the Earth's mantle does not, and nor would a liquid iron core.

Dynamo Theory proposes a mechanism by which a magnetic field is continuously generated. The theory was originally applied to the Sun and other stars, and after to the Earth.

The type of dynamo described in astrophysics and geophysics is the hydromagnetic dynamo. The dynamo requires:

- An electrically conductive fluid medium.
- Kinetic energy provided by planetary or star rotation.
- An internal energy source to drive convective motions within the fluid.

In the case of the Earth, the electrically conductive fluid is supplied by the Earth's liquid metallic core. The kinetic energy is supplied by the Coriolis force which results from the rotation of the Earth. The internal energy source which would drive convective motions within the liquid core is a source of debate. *Encyclopedia Britannica 2011 Standard Edition*: Geomagnetic field: The geomagnetic dynamo – gives thermal heating in the core as the energy source. At one time, it was thought that radioactive elements dissolved in the liquid core produced heat. The present belief is that the centre of Earth's core is a solid sphere of iron. It is proposed that the liquid core freezes onto the outer surface of the solid core; and the energy released in the freezing process heats up the surrounding liquid core.

Zeilik (2002, page 155) explains that there are organized fluid motions within the Earth's core whereby hotter materials rise and cooler materials sink within the hot, liquid core producing convective flows (see footnote [23]). The non-uniform rotation of the liquid core with inner regions carried round faster than the outer region causes convective regions to curl around rather than flowing up and down; these motions are thought to generate magnetic fields.

The Coriolis force is the force felt by a fluid in or on a rotating body. The Coriolis force produced by the rotation of the Earth is thought to introduce twists into the fluid motions within the liquid core. The Coriolis force produces cyclonic storms in Earth's atmosphere.

In order to maintain the magnetic field from a non-permanent magnet such as the one proposed in Dynamo Theory, the convection mechanism must be kept running otherwise the field would decay within 20 000 years. The anomalously long-lived magnetic fields of stars such as the Sun and planets such as the Earth have not been fully explained since numerical models have come up against problems.

The article in *Encyclopedia Britannica 2011 Standard Edition*: Geomagnetic field: The geomagnetic dynamo – claims that the Earth's dynamo mechanism generates electric current by the fluid motion in the core moving conducting material across an existing magnetic field.

---

[23] Convection occurs in liquids under certain circumstances. If heat is generated too rapidly for conduction to carry the heat away, then convection starts to occur to carry away heat. In convection, energy is transported through the liquid by bubbles of hot fluid that rise toward cooler regions, carrying more heat than flows through the same material at rest.

The electric current produces a magnetic field that also interacts with the fluid motion to create a secondary magnetic field with the same orientation as the original field. The two fields together are stronger than the original one. The additional energy in the amplified field comes at the expense of a decrease in energy in the fluid motion.

The article states that a magnetic field must already be present for the geodynamo to work. It is suggested that the existing magnetic field is a relict of the initial formation of planet Earth.

In the next three sections of this chapter I am going to explore how a new model of Earth's structure and formation would produce a geodynamo mechanism. I presented a new hypothesis of Earth's formation in Chapters 17 and 18 with the Planet Capture Hypothesis and in Chapter 21 with the Earth Shrinkage Hypothesis.

4. Independence of core and mantle

There is another curious thing about the Earth's magnetic axis – apart from being upside down; it does not coincide with the rotational axis. The magnetic axis of the Earth is tilted at 11 degrees to the spin axis, and it does not pass exactly through the centre of the Earth. The total field map of the Earth's magnetic field shows that the poles of the dipole are located roughly in northern Canada and on the east coast of Antarctica rather than at the geographic poles. That the field is not exactly centred in the Earth is shown by the field strength not being constant along the equator.

The magnetic axes of the giant gaseous planets have also been found not to be aligned with their spin axes, and often by a large amount.

Jupiter has a magnetic field 10 times stronger than the Earth's with an axis tilted at about 10 degrees from the rotational axis. The polarity of the field is opposite to that of the Earth with the field lines emerging from Jupiter's northern hemisphere and re-entering the planet in the southern hemisphere.

Uranus spins on its side lying almost in the plane of its orbit which means that each geographic pole has 42 years of day followed by 42 years of night. The magnetic field of Uranus is tilted at 59 degrees with respect to the rotational axis; the north magnetic pole is closest to the south geographic pole, and the magnetic field is not centred on the core.

Neptune's magnetic field has about the same strength as that of the Earth. Neptune's magnetic axis is tilted at 47 degrees from the rotational axis and the dipole is offset from the centre.

Saturn's magnetic axis is aligned within 1 degree of the rotational axis with the centre of the field at the centre of the planet. For Saturn magnetic north corresponds to geographic north.

The non-alignment of the magnetic axis of the Earth with the rotational axis has been highly puzzling and no satisfactory explanation has been given up to now. I am going to leave aside the cases of the giant gaseous planets for the moment and suggest an explanation for this phenomenon on Earth.

In Chapter 17 I suggested that planets get their spin from the impetus of the supernova explosion out of which they were born as molten iron planetary cores. In Chapter 18 I suggested that a mantle would form around a planetary core by accretion as the planetary core moved through interstellar dust clouds. The accreted mantle would rotate with the core.

I suggest that the reason why the magnetic axis of the Earth does not align with the spin axis is that the Earth's core has taken on a different spin axis to that of the mantle with its overlying crust.

The description of the Earth's structure in the next section will make it clear why the Earth's core and mantle can move independently.

5. D prime prime layer and cast iron analogy

In this section I am going to take a closer look at the structure of the Earth's core and mantle.

D" layer

Just above the boundary between Earth's liquid core and solid mantle – the core-mantle boundary or CMB – there is what is known as the D prime prime or D" layer. This odd name comes from a classification of Earth's structure proposed by Keith Bullen in 1942 using alphabetical letters. The lower mantle was labelled D, so when two new lower layers were found, they were labelled D prime and D prime prime.

The D prime prime layer is 200 km thick and it has ultra-low velocity zones (ULVZs). These zones were discovered in the 1990s. The ultra-low velocity zones are areas where seismic waves travel slower than they do in the rest of the mantle, and they indicate areas that are in a partially liquid state rather than a solid state.

Sebastian Rost and Justin Revenaugh of the University of California, Santa Cruz have studied seismic shear waves which cannot travel through a fluid and are reflected off the core-mantle boundary. They have detected rigid material within the molten metal of the outer core. They have identified this core-rigidity zone as overlapping the ultra-low velocity zone of the D prime prime layer. The rigid zone detected was a few miles (kilometres) across and about 150 metres thick (See footnote [24] for references).

The areas of low seismic velocity where seismic waves are slowed down by 10 % have high conductivity which suggests the presence of iron. This would link the D prime prime layer to the core. Thus, at the interface between the core and the mantle there appears to exist a

---

[24] Rost, Sebastian & Justin Revenaugh (2001) Seismic Detection of Rigid Zones at the Top of the Core. *Science* Vol. 294: 1911-1914. 30th November 2001.

Geotimes - Redefining the Core-Mantle boundary
www.geotimes.org/jan01/earthsinterior.html
Mineral phase change at the boundary
www.aip.org/pnu/2004/split/679-1.html

comparatively thin layer consisting of solid areas and partially liquid areas.

Some have attributed the ULVZs to perovskite minerals ($MgSiO_3$) of the lower mantle undergoing a phase transition and melting. Others have linked the phenomenon to the core and proposed a hypothesis of Core Sedimentation in which silicates in the core rise to the surface forming solid sediments among areas of liquid iron. Some link this to solidification of the inner core.

Although the D prime prime layer is classified as part of the mantle, it may, in fact, be part of the core.

Cast iron analogy
With the Planet Capture Hypothesis I proposed that the Earth's iron core was smelted within the furnace of a red giant star before being shot out; this model of the Earth's formation and the analogy of cast iron provide an explanation for the D prime prime layer.

The composition of the Earth's core is thought to consist of mainly iron with 4-5 % nickel, 1.5 % chromium, 2-4 % carbon, 2-4 % iron carbide (Fe_3C), >2 % silicon and traces of magnesium, manganese, sulphur and phosphorus – hence resembling a type of cast iron alloy.

In casting iron, nickel may be added to toughen the iron, and if the carbon is removed, it then becomes known as steel. Magnesium may be added to make iron more malleable. Therefore, cast iron includes various types of iron and steel.

When iron ore is smelted in a blast furnace, the metal melts at 1538 $^{\circ}$C (at the atmospheric pressure of the Earth's surface). The ore contains waste rock which also melts and forms slag. Slag consists of molten oxides of silicon, aluminium, sulphur and phosphorus as well as ash. Being lighter than iron, slag forms a liquid layer floating on top of the molten iron. The molten metal is tapped off the bottom of the furnace beneath the slag, and the slag is periodically drawn off to remove it from the iron.

In our analogy, finely divided iron and silicate dust from an interstellar dust cloud are drawn into a red giant star. The globule of molten iron which became the Earth's core was smelted within the star furnace and the lighter silicate material also became molten and separated out from the iron. The molten silicate material was shot out of the star as parent bodies which when captured fragmented into smaller pieces known as meteorites – some meteorites are stony and some are stony-irons. Thus, iron and silicates have mostly been separated out within the star, and the molten iron globule which becomes a planetary core is as pure as cast iron with not more than 10 % impurities.

When a pure metal cools down, the solidification front is very well defined with a clearly delineated solid-liquid interface. The melting point of the pure metal marks the phase change between liquid and solid. However, when a cast iron alloy cools down, the phase change has an upper and a lower temperature called the liquidus and solidus temperatures which give the melting range of the alloy. Above the liquidus temperature the alloy is entirely liquid, and below the solidus temperature, the alloy is entirely solid. Solidification of an iron alloy is characterized by a very thin solidified skin and a large 'mushy zone'. The 'mushy zone' can be called the 'liquid plus solid range' [25].

As the molten planetary core moved through the coldness of space, the cooling effect may have caused it to form a skin of solid iron around the outside. When the core accreted a mantle of silicates from interstellar dust clouds, the skin of solid iron would have prevented the silicates from mixing with the molten liquid core. This would account for the sharp definition of the core-mantle boundary.

Moulds for cast iron are made from sand with a bonding agent (sand is composed of quartz or silica). When cast iron is poured into a mould, the iron solidifies, while the mould made of sand does not melt. In the same way, it is to be expected that when Earth's core accreted a silicate mantle around itself, the surface of the core would have undergone solidification as a thin skin, while the mantle silicates surrounding the core would be heated by contact with the hot metal, but would not melt.

The cast iron analogy suggests that the D prime prime layer belonging to the outer edge of the core containing solid areas represents a skin of

---

[25] Shrinkage in Nodular Iron The Ductile Iron News Issue No.3 2001 www.ductile.org/magazine/2001

solid iron like that around a cooling iron casting. The areas where seismic waves are slowed down by 10 % would represent a 'large mushy zone'. The mushy zone of cast iron contains crystals of solid iron amongst liquid molten iron. The area is 'mushy' because the various metals and other elements in the alloy produce different liquidus and solidus temperatures which make solidification uneven.

Therefore, my conclusion is that the ultra low velocity zones of the D prime prime layer which are partially liquid and partially solid, and have high conductivity typical of metals are areas around the edge of the Earth's core which have undergone cooling and crystallization. This type of solidification is typical of cast iron alloys, and we know the Earth's core to be an iron alloy which is cooling down.

The model of the interior of the Earth which I am proposing represents the core as being entirely molten except for a solid skin around its edges like cooling cast iron. In this model there is no solid inner core.

According to this model, the heat of the Earth's core is attributed to its mode of formation forged within a red giant star. This is as opposed to the current theory which attributes the interior heat of the Earth to kinetic energy produced by impactors hitting the outside of the planet. If the Earth's heat originated deep inside the planet, and did not come from an outside heating method, one would expect the middle – the inner core to be the hottest part of the planet.

If the centre of the core is the hottest part of the planet, then any solidification that occurred would happen around the edges of the core working progressively towards the centre, and not from the centre outwards.

Earth's permanent magnet
According to current theory the inner part of the Earth's core is solid. The atmospheric pressure of the Earth's inner core is between 3.3 and 3.6 million atmospheres. At this high pressure iron forms a solid at 5000 °C (Lamb & Sington 1998, page 101). It is for this reason that the Earth's core is thought to have a temperature of between 5000 and 6000 °C.

I do not believe that the inner core is solid. For the whole core to be molten liquid except for an outer solid layer at this pressure, it must have a temperature above 6000 °C. In Chapter 18, section 3 I suggested that the core may have a temperature as high as 30 000 °C (see footnote 26).

If the Earth's core is cooling like a lump of cast iron with a zone of solidified iron around it, the solid iron skin must have a temperature of 5000 °C or less to be solid at the pressure of the interior of the Earth.

Just as iron can be solid at higher temperature when it is under higher pressure, I am assuming that the Curie point occurs at a higher temperature deep inside the Earth. Where the pressure is very high, the Curie point may not be restricted to about 770 °C as at one atmosphere pressure on the Earth's surface.

The importance of this is that, a thin solid outer layer of the Earth's core may act as a permanent magnet producing a primary magnetic field.

6. Earth's dynamo – a new model

In this section I am going to present a new model for the working of Earth's dynamo with an explanation for reversals in Earth's magnetic field. First, a note on the workings of dynamos.

Dynamos

A dynamo contains a magnet and a wire coil. When the magnet is spun round by mechanical force it produces a changing magnetic field that causes electrons in the wire to be set in motion. The result is an electric current produced in the wire coil induced by electromagnetic

---

26 The Earth is losing heat at the surface. The temperature increases by 10-15 °C within the crust for every kilometre down. If the mantle loses heat by conduction (and not convection) then the core may have a temperature of between 30 000 and 40 000 K which may correspond to the heat inside a red giant star. This temperature for the Earth's core would exclude solidification at its centre.

induction. It does not matter whether the magnet or the coil is moving, so long as they move relative to each other.

Electricity generators are dynamos. They use spinning magnets to produce electric current. There are direct current (DC) generators and alternating current (AC) or synchronous generators. In AC generators called alternators, the current reverses polarity at a fixed frequency of usually 50 or 60 cycles or double reversals per second.

Generators can use field windings or permanent magnets mounted on a ring to produce a magnetic field. Permanent-magnet generators require no provision of field current; however, the voltage varies with changes in load.

The liquid core of the Earth is thought to act as an electromagnet generating shifting electric currents and thereby creating a magnetic field.

Earth's dynamo
This is a proposal for a new model of Earth's geodynamo which accords with the new way of understanding Earth's formation presented in Chapters 17, 18 and 21:

The Earth's core is losing heat and cooling down. As it cools, liquid iron contracts in volume but the shrinkage is much more acute when iron solidifies. If the Earth's core does indeed have a solid layer of iron surrounding it, shrinkage will have caused the core to pull away from the surrounding mantle.

The mantle composed of silicate minerals is hot, but solid due to the immense pressure it is under. However, shrinkage in the volume of the core would release pressure leading to decompression of the lower layer of the mantle in contact with it. Decompression would lead to a layer of molten magma forming between the solid iron outer layer of the core and the solid layers of the mantle.

The core shrinking away from the mantle and the layer of molten magma formed would allow the core and mantle to move independently. Slippage may occur between the core and mantle. Random slippage would lead to the core rotating at a slightly different angle to the mantle. The Earth's core and mantle spin on axes with

11 $^\circ$ difference as indicated by the poles of the magnetic field and the geographic poles.

The Earth's dynamo resembles a direct current generator with permanent magnets to produce the magnetic field. I am proposing that a solid skin of iron around the core (which caused the core to shrink away from the mantle since solid iron shrinks more than liquid iron) would act as a permanent magnet due to ferromagnetism. This would be so if the Curie point occurs at a higher temperature in the Earth's interior where the pressure is higher than at the surface of the Earth. This permanent magnet the size of the circumference of the core (although broken in places) would produce a magnetic field.

Shifting electric currents produced within the liquid core would act as moving charges producing magnetic flux. The shifting electric currents would be produced within the highly conductive liquid metal of the core by the Coriolis forces of the rotating planet. In addition, the differential in rotation of the core and mantle caused by slippage may augment the production of electric currents.

Consequently, the spinning liquid core would produce electric currents converted into magnetic flux, while the solid outer parts of the core forming a skin broken in places would act as permanent magnets providing a magnetic field. This set-up would drive the magnetic field emanating from and re-entering the poles of the core.

Reversals and battery power
As noted in section 1 of this chapter, the Earth's magnetic field undergoes periodic reversals in polarity documented in volcanic and sometimes sedimentary rocks belonging to the geological record.

After much searching, it occurred to me that the reason for reversals in Earth's magnetic field is because the operational system of electric current production in Earth's core undergoes periodic changes in mode. To understand this we need to understand how a battery operates.

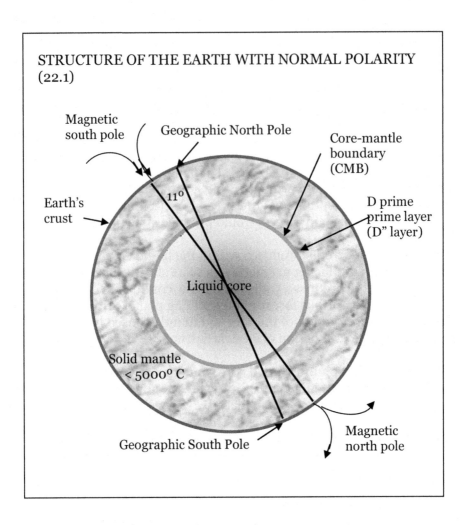

STRUCTURE OF THE EARTH WITH NORMAL POLARITY (22.1)

Batteries convert chemical energy into electrical energy. A battery has a positive cathode and negative anode. These electrodes are separated by and often immersed in an electrolyte which permits the passage of ions between the two electrodes. An electrode contains only a limited number of units of chemical energy convertible to electrical energy; therefore a battery of a given size has a certain capacity and will eventually become exhausted.

The electrolyte of a battery consists of a solvent such as water, although it could be a solid, and one or more chemicals that dissociate into ions in the solvent. These ions serve to deliver electrons and chemical matter through the cell interior balancing the flow of electric current to the outside during cell operation. Battery performance depends on the diffusion rates of the internal chemicals through the electrolyte solution.

In 1834 Michael Faraday determined that for every ampere of electric current that flows for a period of time, a matching chemical reaction or other change must take place. This is the principle upon which batteries operate.

Differential rotation: I have proposed that the Earth's core and mantle rotate on slightly different axes, currently 11 degrees difference. The differential rotation would produce friction which would generate heat. Heat would accumulate at the junction between the core and mantle until the thin solid skin of iron around the core would start to melt and break down.

In places where the solid skin has broken down, molten magma from the mantle would break through into the core.

When this occurred, the remaining solid parts of the skin would suddenly start to act as iron electrodes conducting electricity in the mode of a battery. The liquid molten magma fed into the liquid core would produce ions. The Earth would produce electric current in the mode of a battery with the solid parts of the skin around the core acting as negative iron anodes, the magma acting as positive cathodes providing ions and the liquid metal core acting as a nickel hydroxide electrolyte.

The change from the mode of a direct-current generator with the solid skin acting as permanent magnets to the mode of a battery with the solid parts of the skin acting as negatively charged electrodes would cause a sudden reversal in polarity in Earth's magnetic field.

Disruption of the solid skin around the core would switch the battery on. When the molten magma providing ions ran out, the battery would be switched off. With a reduction in heat, the almost complete solid skin around the core would reform again and the interior dynamo would resume electricity production. This mechanism would switch normal and reversed polarity on and off.

The question is, which polarity would battery power or dynamo power correspond to? My guess is that we would be running on battery power now. If this is the case, it would mean that normal polarity corresponded to battery power, while reversed polarity (aligned with the geographic poles) would be driven by dynamo power involving shifting electric currents within the core. Signs that we have been running on battery power through the Cenozoic era are that the battery appears to be running low at the moment and the strength of the output constantly wavers.

Wavering
The magnetic field of the Earth wavers in its direction and strength. Wavering is measured as 'secular variation' which is a map of the varying strengths of the magnetic field over the Earth's surface. The origin of the secular variation is not known.

Secular variation shows a westward drift of magnetic anomalies. *Encyclopedia Britannica*: Geomagnetic field: Secular variation of the main field – states "The westward drift of magnetic anomalies evident in the secular variation should provide an important clue to the origin of the main field if only it can be interpreted." The Earth rotates eastward. If features within the core rotate more slowly than the surface features, they will appear to move backward relative to the general rotation i.e. westward.

In section 5 of this chapter I described the D prime prime layer located at the core-mantle boundary which I identified as a solidifying part of the liquid core. The analogy of cast iron showed that it is to be expected that a molten liquid iron alloy core in the process of cooling down would have an outer thin solid skin of iron with an underlying 'mushy zone'. In the case of a cooling planetary core the mushy zone would consist of slabs of crystallizing iron alloy resembling large pancakes of solid iron surrounded by liquid iron.

I have proposed that around the outer edges of the core, solid iron reaches a low enough temperature within the high pressure conditions of the interior of the Earth to reach the Curie point and become ferromagnetic. Iron in its solid state becomes magnetic by the iron atoms of its domains being organized to all point in the same direction. Heating disorganizes the domains which is why heating iron destroys its magnetic properties.

248

Variations in temperature at the core-mantle boundary would cause the outer parts of the core to solidify into solid 'pancakes' and melt again with the gain and loss of magnetic properties. This would produce an intermittent magnetic field which would explain changes in the strength of Earth's magnetic field.

It is likely that a partially liquid mushy zone would move round slower with the rotation of the Earth than a solid outer skin. This would cause the magnetic signatures produced by solid pancakes of iron within the mushy zone to show a westward drift.

Iron crystals below the Curie point channel electromagnetic currents in accordance with the orientation of the crystals. If crystals channelling magnetic field lines on a north-south dipole were displaced by westward drift, this would produce the magnetic signatures observed involving changes of axis of the magnetic field at different points on the Earth's surface.

Another reason for the wavering of the Earth's magnetic field may involve magnetic permeability. Ferromagnetic materials such as iron do not have a constant relative permeability. As the magnetizing field increases, the relative permeability increases, reaches a maximum, and then decreases (*Encyclopedia Britannica 2011 Standard Edition*: Magnetic permeability).

Magnetic fields do not remain constant. A permanent magnet or a wire carrying a steady electric current in one direction will produce a stationary magnetic field in which the magnitude and direction remain the same at any given point. However, around an alternating current or a fluctuating direct current, the magnetic field is continuously changing its magnitude and direction (*Encyclopedia Britannica 2011 Standard Edition*: Magnetic field).

I have likened the Earth's dynamo to a direct current generator with permanent magnets creating a field. The direct currents produced in the Earth's core are likely to be fluctuating. This would be another source of wavering in Earth's magnetic field.

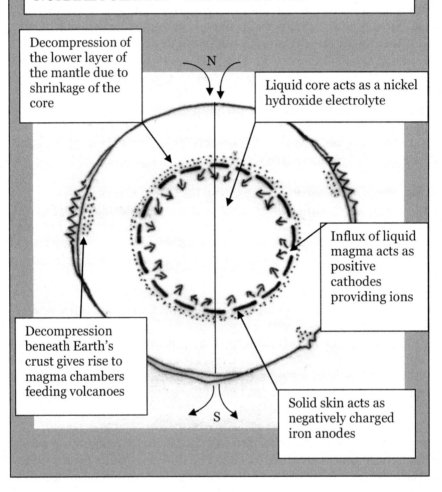

22.2 NEW MODEL OF EARTH'S INTERIOR SHOWING THE GENERATION OF EARTH'S MAGNETIC FIELD AND REVERSALS IN POLARITY

NORMAL POLARITY – BATTERY POWER

Decompression of the lower layer of the mantle due to shrinkage of the core

Liquid core acts as a nickel hydroxide electrolyte

N

Influx of liquid magma acts as positive cathodes providing ions

Decompression beneath Earth's crust gives rise to magma chambers feeding volcanoes

S

Solid skin acts as negatively charged iron anodes

250

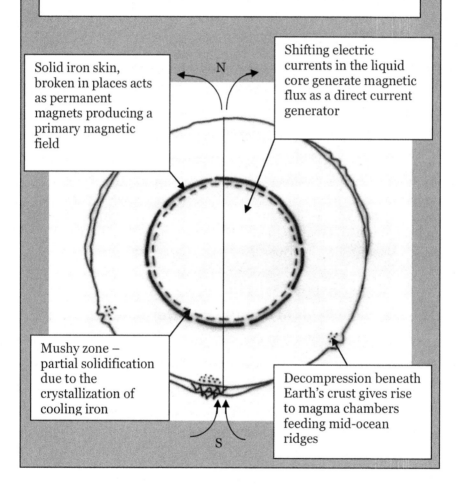

22.3 NEW MODEL OF EARTH'S INTERIOR SHOWING THE GENERATION OF EARTH'S MAGNETIC FIELD AND REVERSALS IN POLARITY

REVERSED POLARITY – DYNAMO POWER

Solid iron skin, broken in places acts as permanent magnets producing a primary magnetic field

Shifting electric currents in the liquid core generate magnetic flux as a direct current generator

Mushy zone – partial solidification due to the crystallization of cooling iron

Decompression beneath Earth's crust gives rise to magma chambers feeding mid-ocean ridges

N

S

Changes in Earth's magnetic field over Earth history

In terms of the history of Earth's magnetic field, the cast iron analogy suggests that if Earth passed through a stage of being a denuded planetary core travelling through space, it would form a solid skin around the hot molten body of the core through being exposed to the cold conditions and low pressure conditions of space.

The formation of a thick solid skin of iron completely encircling the core may preclude the existence of a geodynamo producing a magnetic field at this stage.

The accretion of a thick mantle in its proto-planet stage would cause an insulating effect preventing immediate heat loss into space from the core and produce a sharp rise in pressure in the Earth's interior. Under the atmospheric pressure of the Earth's surface iron solidifies at 1538 °C, while under the high pressure conditions of the Earth's core iron solidifies at <5000 °C.

The interplay between temperature and pressure at the core-mantle boundary may have caused a thick solid skin to re-melt in places or remain only as a thin solid skin around the core. This would create a complex picture and may help to explain the changes in Earth's magnetic field over geological time.

7. The magnetic fields of other terrestrial planets

Do the other terrestrial planets of the solar system possess dynamo mechanisms for producing magnetic fields like Earth?

Mercury has a weak geomagnetic field; Mars has palaeomagnetism but no currently generated magnetic field; and Venus has no detectable magnetic field from an internal source.

Mercury

The magnetic field of Mercury is only 1 % as strong as that of the Earth and is nearly orientated along the planet's axis of rotation. It is possible that Mercury's magnetic field is produced by remanent magnetism alone; however, it is thought that Mercury still possesses a dynamo mechanism.

An indication that Mercury has a molten liquid iron core is that the surface of Mercury shows signs of shrinkage (see Chapter 21, section 3). This is likely to be produced by shrinkage of the core. Shrinkage is more likely to occur when molten iron cools and reduces in volume or when molten iron solidifies, although solid iron also reduces in volume as the temperature lowers.

If Mercury has a molten iron core, there is the possibility that it generates electric currents that produce a magnetic field in the same way as Earth. The weakness of Mercury's magnetic field may be due to its very slow rotation of 59 days to complete one rotation with its rotation locked onto the rotation of the Sun. This slow rotation would drastically reduce the Coriolis force.

Mercury's slow rotation may not be enough to drive the generation of electric currents in the core to produce a magnetic field as strong as Earth's magnetic field. It is also possible that Mercury's core may have solid iron forming around its edges as it cools, and that this solid layer of iron may simply be acting as a permanent magnet producing Mercury's weak magnetic field. A considerably thick skin of solid iron completely encasing the core could prevent a dynamo mechanism from producing magnetic flux beyond the confines of the core.

Venus

Venus has a core similar in size to Earth's core. It is not known whether the core of Venus is liquid, solid or partially liquid. An indication that Venus' core may still be partially molten liquid metal is that like Mercury, the surface of Venus shows signs of shrinkage.

Venus barely rotates at all compared to Earth since it takes Venus 243 Earth days to complete one rotation. This very slow rotation may not be enough to drive a dynamo mechanism to produce a magnetic field. The absence of a magnetic field around Venus means that the solar wind runs straight into Venus causing turbulence in its atmosphere.

There remains the question of why Venus does not have any magnetic field since remanent magnetism could produce a weak field.

We saw in Chapter 17 that everything about Venus is at odds with the other terrestrial planets in the solar system – I wonder whether Venus was captured from another galaxy and thus has a different composition to that of other terrestrial planets in the solar system. If the core was composed of antiferromagnetic materials (see section 2 of this chapter), then solidification of the core of Venus would not produce a magnetic field through permanent magnetism.

Mars

Mars does not show evidence of current production of a magnetic field, but parts of its crust show remanent magnetism. The palaeomagnetic signatures of Mars' crustal rocks show that alternating polarity reversals of its dipole field have occurred in the past. Mars is thought to have had a planetary dynamo 4000 million years ago.

Mars has a core composed of iron and nickel with about 16-17 % sulphur. This represents twice the concentration of lighter elements that exist in the Earth's core. Mars' iron sulphide core is thought to be partially fluid. It may be that the composition of Mars core with a higher percentage of lighter non-magnetic elements has caused Mars to lose the dynamo producing its former magnetic field.

8. Dynamos of the Sun and giant gaseous planets

We saw in Chapter 13 that the giant gaseous planets in the solar system – Jupiter, Saturn, Uranus and Neptune – may be failed stars which have cooled down and shrunk in size. Their identity as stars is possibly confirmed by their production of magnetic fields in a manner similar to that of the Sun.

The Sun, other stars and the giant gaseous planets are composed mainly of hydrogen with the consistency of water. Hydrogen behaves as a liquid rather than a gas at pressures above 1000 atmospheres. The Sun's core has a pressure of 250 000 million atmospheres and at the centre of the core the density is over 150 times that of water. The pressure inside Jupiter reaches two million atmospheres and the pressure inside Saturn is well over a million atmospheres.

The following description of how magnetohydrodynamic power generators (MHD) work provides some insights into the possible workings of the Sun's generation of electricity and magnetic fields emerging from sunspots on its surface, as well as the generation of magnetic fields by the gaseous planets.

Magnetohydrodynamic power generators (MHD)

Magnetohydrodynamic power generators generate electric current by the interaction of an electrically conducting fluid and a magnetic field. The fluid is driven by a primary energy source such as the combustion of coal or gas through a magnetic field. This produces voltage in the conductor.

MHD generators working at lower temperatures may use liquid metal as the electrically conducting medium, while MHD generators which work at high temperatures of 1800 K upwards use liquid hydrogen. When nuclear heat is employed as the primary energy source, hydrogen is used as the working fluid.

Liquid hydrogen gas forms plasma in which free electrons are available to be organized into an electric current in response to an applied or induced electric field. The plasma must also contain positively charged ions to balance the electric charge. The plasma will contain some neutral atoms or molecules.

For MHD generators to operate competitively adequate conductivity of the plasma is maintained by injecting a small amount of additive – typically about 1 % by mass – into the working fluid of the MHD system. The additive consists of readily ionisable material such as potassium carbonate and is referred to as "seed". The additive is the principal source of electrons and ions that render the plasma electrically conducting.

The power output of MHD generators for every cubic metre of conductor depends directly on its conductivity, the square of the velocity at which the conductor moves, and the square of the magnetic field through which it is passing (*Encyclopedia Britannica 2011*

Standard Edition: Electric generator: Magnetohydrodynamic power generator).

Star dynamos

Stars are composed of hydrogen plasma which makes them good candidates for generating magnetic fields with a type of hydromagnetic dynamo.

Unlike MHD generators, however, the Sun and gaseous planets are unlikely to possess magnetic fields produced by the ferromagnetism of permanent magnets. The magnetic fields of stars or failed stars are more likely to be produced by moving electric charges.

The differential in the rotation of their gaseous atmospheres may produce electric charges moving relative to each other giving rise to magnetic fields. The outer layers of the Sun exhibit differential rotation: at the Sun's equator the surface completes one rotation in 25.4 days, while near the Sun's poles it takes 36 days to complete one rotation; meanwhile the Sun's core rotates as a solid body.

A similar phenomenon occurs with the gas planets too; since they consist of gas rather than rocky surfaces and mantles, their atmospheres can rotate at slightly different velocities between the equator and poles. The giant gaseous planets all rotate quite rapidly. Jupiter has a rotational period of 9 hours, 55 minutes and 30 seconds. The rotation of Jupiter's magnetic field is taken as the rotation of Jupiter as a body. Saturn has a rotational period of 10 hours and 14 minutes. Uranus has a rotational period of 17 hours at the equator. Neptune has a rotational period of 16 hours and 3 minutes.

As already noted, magnetic fields produced by moving electric charges are orientated perpendicular or the angle between the thumb and forefinger to the electric current flow.

Engulfed bodies: It is interesting that in MHD generators, electric current is generated by plasma with a seed of ionisable material thrown into it.

The Sun engulfs any body in the solar system that strays too near to it. Indeed the Planet Capture Hypothesis asserts that the Sun is constantly capturing bodies from outside the solar system, and that many of these disappear into it. Bodies such as comets, meteorites, asteroids and even larger bodies engulfed by the Sun would provide rocky material that would dissociate into ions within the Sun's plasma, raising its conductivity.

The gaseous planets also draw in bodies such as comets, meteorites and asteroids. Some moons will eventually collide with and become engulfed by the gas giant they currently orbit. The Great Red Spot in Jupiter's atmosphere may mark one such collision location.

Do sunspots represent the locations where 'seed' was flung into the plasma generator of the Sun?

There is the possibility that sunspots mark the locations where 'seed' has been flung into the Sun's plasma with the energy re-emerging as magnetic flux. If this were the case, solar flares would represent occasional belches of energy from the digestion of a larger than average rocky meal.

If sunspots and solar flares were produced by small bodies engulfed by the Sun, why would sunspots follow a cycle of increasing incidence over 11 years and a return to original polarity after 22 years?

The Sun orbits the centre of the Milky Way clockwise as viewed from the galactic north pole in about 225-250 million years at a speed of about 251 km/second. The Sun is currently travelling through the Local Interstellar Cloud in the Local Bubble zone within the inner ring of the Orion arm of the Milky Way.

The Milky Way is also part of the orbiting of the Local Group – a group of over 20 galaxies to which it belongs.

The Local Bubble is a region of hot rarefied gas within our galaxy. It has been argued that the Sun's passage through the higher density spiral arms coincides with mass extinctions on Earth, due to increased impact events.

If this orbiting brought the Sun through regions of space containing a higher density of bodies on an 11 year or 22 year cycle, and the Sun drew in and engulfed a higher number of bodies while passing through these regions, this could explain the periodicity in the incidence of sunspots.

There is the possibility that the capturing of rocky bodies of varying size by the Sun keeps its hydromagnetic dynamo primed for the production of magnetic fields which emerge from sunspots and become stronger during solar flares.

Conclusion

In this final chapter we have seen that the Earth's magnetic field emanating from the core is characterized by occasional polarity reversals over geological time; constant wavering in strength and an oscillation in direction; it is currently non-aligned with the axis of Earth's geographic poles by 11°.

The Sun's magnetic field emerges from sunspots; is strongest when solar flares occur; and it engulfs the solar system, flowing past the Earth as solar wind. The Earth is protected from the accelerated particles of the solar wind by its magnetosphere produced by Earth's magnetic field.

Magnetic fields may be produced by permanent magnets which are composed of ferromagnetic or iron-rich materials, or by induction from moving electric charges.

The work of William Gilbert published in 1600 led to the conclusion that the Earth's primary magnetic field is the result of permanent magnetism as if the Earth contained a giant bar magnet. In the 20th century this notion was rejected in favour of the proposal that the Earth has an induced field produced by electric currents in a fluid core.

Dynamo Theory involves complex explanations for the production of electric currents in Earth's core. However, the source of energy for a self-sustaining induced magnetic field has not been determined. All current proposals come up against the problem that the dynamo mechanism would die away after a comparatively short time counted in thousands of years. I think that the idea that the Earth has a magnetic field initially produced by permanent magnetism could be resurrected with some additions to the mechanism.

My own conclusion is that the dynamo mechanism producing the magnetic field of the Earth is different from the dynamo mechanism producing magnetic fields around stars such as the Sun or failed stars such as the planets Jupiter, Saturn, Uranus and Neptune. Just as dynamos or generators producing electricity are of various types, I believe that geodynamos are also of various types.

I have proposed that the Earth's geodynamo can be likened to a direct-current generator, with conducting fluid producing shifting electric currents and permanent magnets producing the field as one mode of operation. The dynamo mechanism also operates on another mode likened to that of a battery, involving chemical energy and the transport of ions through a conducting fluid between electrodes. Reversals in the polarity of Earth's magnetic field would correspond to suddenly occurring changes in the mode of operation.

I have proposed that normal polarity (with the north magnetic pole oscillating around the south geographic pole) corresponds to battery power which will eventually run out, and that reversed polarity in which the magnetic axis points in the same direction as the geographic axis corresponds to direct-current dynamo power.

I have likened the dynamos producing magnetic fields around the Sun and gaseous planets to magnetohydrodynamic power generators (MHD) which function with plasma in the form of liquid hydrogen.

I have not presented these ideas concerning geodynamos as another hypothesis, but as predictions of other hypotheses I have proposed in Part III.

The notion of shrinkage of Earth's core leading to independence of the core and mantle is a corollary of the Earth Shrinkage Hypothesis. This involves the idea that between the solid skin of a cooling, shrinking core and the solid mantle, decompression has led to melting of the lower layer of the mantle. This liquid layer of mantle would allow slippage to occur between the rotational axis of the core and the rotational axis of the mantle. The evidence for this is the 11° non-alignment of Earth's magnetic field axis with the geographic axis.

The differential in rotational axes may contribute to the dynamo mechanism of the core in the production of electric currents. It may also cause friction producing heat which would lead to a break-down of the solid iron skin around the core. This would allow the influx of molten liquid mantle into the liquid core; this would change the mode of operation of the geodynamo.

Overall this model of Earth's core and the proposed dynamo mechanisms constitute a prediction of the Planet Capture Hypothesis. The Planet Capture Hypothesis – which involves the notion of planetary cores shot out of supernova explosions and travelling through interstellar space until captured by a star – presents the Earth's core as entirely molten liquid iron alloy at a much higher temperature than supposed up to now.

Since the heat in Earth's core would have come from the red giant star in which it was forged, the analogy of cast iron is applicable. The cast iron analogy shows that the iron core cooling down would form a solid skin around the outer edges with a 'mushy zone' below this.

On the assumption that solid iron at high pressure in the interior of the Earth can be ferromagnetic, the permanent magnetism of the solid iron layer around the core and solid pancakes in a mushy zone would help to explain many features of Earth's magnetic field in terms of its wavering.

A further prediction of the Planet Capture Hypothesis involves the possibility that the Sun's dynamo mechanism is kept primed by rocky bodies captured from outside the solar system being engulfed by the Sun in the manner of 'seed' flung into an MHD generator to maintain its electrical conductivity.

Thus, I end here with four scientifically testable predictions:

- Independence in the rotation of core and mantle of the Earth.

- A solid outer layer of iron around the core which plays a pivotal role in the two dynamo modes of Earth's core.

- The existence of a 'mushy zone' of partial solidification causing wavering in Earth's magnetic field.

- The engulfing of small bodies by the Sun as the cause of sunspots from which the Sun's magnetic field emerges.

My hope is that these predictions will lend support to the Planet Capture Hypothesis and Earth Shrinkage Hypothesis. These hypotheses present a new mode by which planet Earth was formed and the solar system came into existence. This new explanation presented here for the generation of Earth's magnetic field is a simple solution to a complex problem.

Bibliography

Encyclopedia Britannica 2011 Standard Edition: Battery; Chemical element: the Earth's core; Curie point; Earth: the geomagnetic field and magnetosphere, the interior; Electric generator; Energy conversion; Faraday, Michael; Geochronology; Geomagnetic field; Gilbert, William; Iron processing; Mercury: the magnetic field and magnetosphere; Nickel processing; Ohm's Law; Polar wandering; Saturn: the magnetic field and magnetosphere; Steel; Sun.

Ince, Martin (2007) *The Rough Guide to the Earth* Rough Guides Penguin Group

Lamb, Simon & David Sington (1998) *Earth Story: The Shaping of Our World* BBC Books

Zeilik, Michael (2002) *Astronomy: The Evolving Universe* 9th edition Cambridge University Press

CONCLUSION TO PART III

Chapters 17 to 22 which form Part III of *The Steps of Creation* describe natural processes. The three hypotheses of Part III are not part of the Theory of Nanocreation and Entropic Evolution; they do, however, provide a link between life in the universe, life on Earth and the formation of the Earth.

The natural events described in Part III are contingent or fortuitous – they do not happen by design. Their occurrence is governed by chance. Any predictability in their occurrence is only a consequence of a chain of natural events each incident upon the other and upon the generality of their occurrence.

Summary of hypotheses in Part III

The **Planet Capture Hypothesis** (Chapter 17) states that:

The solar system was formed by our star, the Sun capturing travelling proto-planets through gravitational attraction. Planets originated as molten iron planetary cores forged within red giant stars and shot out by nova or supernova explosions. These molten iron cores accreted silicate mantles of cold interstellar dust while travelling through space. Thus, the planets of the solar system are of different ages and origins.

I propose that the solar system is not a self-contained system as the Nebular Theory proposes. The solar system of planets orbiting the Sun is presented in the Planet Capture Hypothesis as being formed principally by the fortuitous capture by the Sun of planets that were travelling through space along with various other objects. The variety in the forms of orbits, rotational axes, rotational speed and composition of the different planets and moons in the solar system constitute strong evidence in support of the Planet Capture Hypothesis.

Comets provide a test-case for the Planet Capture Hypothesis. Comets are 'dirty snowballs' of water ice and dust. Their tails show that they are evaporating away which means that they cannot be ancient occupants of the solar system. There are long-period comets and short-period comets. Long-period comets have highly elliptical orbits around the Sun which take them beyond the furthest planets and their orbits are inclined at all angles to the plane of the solar system, some having retrograde orbits. Short-period comets which return to the inner solar system in less than 200 years have less eccentric orbits that conform to a greater degree to the plane of the solar system, and they all orbit in the same direction as the planets.

It appears to be the case that long-period comets are converted into short-period comets after a certain number of transits through the solar system over thousands of years. I propose that the orbits of comets become less eccentric as they lose energy. Comets would represent recent capture events occurring less than a million years ago; the gradual conformation of their orbits to the solar system demonstrates what happened to the orbits of planets captured a much longer time ago.

The structure of the Earth is described in Chapter 18. The division into core, mantle and crust can be explained by the Earth's mode of formation. Integral to the Planet Capture Hypothesis is the notion that planetary cores originated as molten iron globules smelted in dying stars and shot out by nova or supernova explosions. Accordingly, the heat of Earth's core would come from the heat inside the star which forged it.

The Planet Capture Hypothesis involves the idea that molten iron planetary cores moving through galactic space acquired mantles composed of silicate minerals by accreting material as they passed through the debris of exploding stars, interstellar dust clouds and the planetary disk of the capturing star. The composition of the mantle and the thickness would be dependent on which clouds the planetary core had passed through. A planetary core would become a proto-planet when it had accreted a mantle of minerals, and a proto-planet would become a planet when it had been captured into orbit around a star.

During the journey through space, silicate minerals would build up as a layer surrounding a molten core from accretion of cold, silicate interstellar dust grains. At the same time, water ice and organic carbon molecules associated with interstellar dust would also be accreted. These volatiles would be vaporized and start to form the beginnings of an atmosphere. Unicellular anaerobic bacterial type life could also have been accreted *in situ* from dark clouds onto the surface of the proto-planet Earth when enough mantle had been accreted to provide a cool surface.

The Planet Capture Hypothesis is applicable to all planets and moons in our solar system and to planets in other planetary systems. The same principles can also be applied to smaller bodies such as meteorites.

The **Mass Extinctions on Earth Hypothesis** (Chapter 19) states that:

The mass extinctions on Earth are evidence of cataclysmic events provoked by disturbances in the solar system involving planet capture. Gravitational interaction between the Earth and a captured planet or large body passing near to the Earth have caused changes in the axis of rotation of the Earth leading to ice ages; earth tides leading to episodes of volcanism; and the gravitational pull on the oceans has produced giant tidal waves. Giant tidal waves were also produced by asteroid strikes to the oceans on Earth during certain periods. Deposition of sediments by tidal waves was the main agency of mass extinctions and the concurrent formation of deep fossil beds.

Disturbances in the solar system causing mass extinctions of life on Earth have included, according to the Mass Extinctions on Earth Hypothesis: the capture of planets into the inner solar system crossing the orbit of the Earth to take up an orbit closer to the Sun (Venus, Mercury); planets or large-sized objects passing close by the Earth before being engulfed by the Sun and disappearing; a massive collision of a large body with a planet which once orbited where the asteroid belt is now found, resulting in asteroids sent spinning in all directions crossing the paths of other planets; and collisions of satellite moons with giant gaseous planets in the outer solar system sending shock waves through the solar system.

The Mass Extinctions on Earth Hypothesis lends support to the Planet Capture Hypothesis since it links the evidence of extinctions on Earth to disturbances in the solar system.

We observe that despite evidence of massive disasters in the past, Earth still has abundant life. Other planets are equally likely to have had unicellular life as the Earth; however, it seems that most other planets have undergone ecological disasters that have ended their history as good places to live.

The degree to which life develops on any given planet depends upon the ecological conditions of that planet. The ecological conditions may only allow life to develop as far as Archaea or bacteria. Some planets may have carried life as far as Protista complexity. We only know of our own planet as having developed multicellular kingdoms of life.

Important issues presented in Part III, although not presented as hypotheses include the organic origin of water on Earth and the organic origin of the sedimentary rocks of Earth's crust. Chapters 18 and 20 contain discussions on the formation of the oceans and rock strata of Earth's surface.

In the beginning the Earth had no oceans, no aerobic atmosphere, no sedimentary rocks and no land; it resembled other terrestrial planets in the solar system with a dry basalt surface. The minerals of the mantle formed the first surface.

Anaerobic bacteria were the first inhabitants of Earth to start to change this situation. Archaea and Eubacteria may have colonized the Earth when it was still a proto-planet travelling through interstellar space. It may even have been colonized several times over in between episodes of accretion of the mantle, leaving several ancient surfaces sandwiched into the mantle (now sources of diamonds).

Water
The first water on Earth was produced by the metabolic processes of anaerobic bacteria, drop by drop over geological time scales measured in thousands of millions of years. By the Cambrian period the Earth had become covered in shallow seas.

My conclusion presented in Chapter 18 is that the water *presently* found on Earth comes mainly from the following sources:

- The metabolic activities of metal-oxidizing aerobic bacteria living on basalt in which each metabolic conversion is accompanied by the production of water. There is evidence of the extensive biological weathering of basalt formations of ocean floors, and large accumulations of metal ores in the form of nodules and crusts at the bottom of the oceans.

- The conversion of ancient buried organic matter by bacteria. Oil fields contain anaerobic Archaebacteria called methanogens which convert subsurface accumulations of organic matter into kerogen –a precursor to oil, as well as methane gas and abundant water. The burning of fossil fuels produces carbon dioxide and water. The organic matter of oil fields consists mainly of cyanobacteria and planktonic algae accumulated during the Precambrian era.

- The conversion of subducted marine sediments containing organic matter into gases and water by combustion caused by volcanism. The water is blown out of volcanoes as steam. Also, the conversion of buried organic matter in ocean floor sediments by combustion associated with mid-ocean ridge volcanism. Reduced gases and liquid water escape out of hydrothermal vents close to mid-ocean ridges.

  There may also be subsurface bacteria living on organic matter found in marine sediments, converting it into gases and water in the vicinity of hydrothermal vents where the rocks are warm rather than hot. This may also contribute to the water emerging from vents in the ocean floor.

In Chapter 10 of Part I and again in Chapter 18 of this book I presented the view that the amount of water on Earth is continually rising. The shallow seas of the Cambrian are now deep oceans.

The transformation of basalt

It was oxygen in the atmosphere of the Earth which allowed the true transformation of the planet into a unique place. The availability of oxygen made it possible for iron-oxidizing and manganese-oxidizing aerobic bacteria to live on basalt as lithotrophs –rock-eaters. These lithotrophic bacteria as part of a community of different sorts of bacteria performed various molecular transformations leading to the formation of sedimentary rock. They converted the reduced metals found in basalt into iron oxides and manganese oxides found as crusts and nodules of metal-rich ores on the ocean floor; they converted the minerals in basalt into the secondary minerals of mica and fine clay which contains the elements vital to life; the biological weathering of basalt released elements to seawater, especially sodium which made the sea salty; and the by-product of these metabolic processes was water.

The clay of the seabed became the substrate for microscopic forms of life and worms. Mud formed in the guts of worms has become a major type of sedimentary rock called mudstone or shale. Shale is often metamorphosed into slate.

Therefore, bacterial lithotrophs appear to be the key to the initial decomposition of the minerals found in basalt transforming them into secondary minerals. This may be followed by vermiform transformation of clay into mud which became shale. These transformations allowed elements to be made available to higher forms of life.

Limestone and chert
Protists of the Precambrian living in pools of water (the size of large lakes or small seas) appear to have been responsible for the production of some of the first sedimentary rocks. The accumulation of new minerals on Earth occurred through the secretion by these mainly unicellular organisms of shells of silica or calcium carbonate for protection and support.

The deposition of limestone and bedded chert on sea floors appears to coincide with the advent of protist forms of life such as foraminifera, algae, diatoms and radiolaria some 2.5 thousand million years ago. Photosynthetic plankton consisting of protist motile algae lived as huge blooms in surface waters, while protozoan scavengers lived on the seabed. The seasonal die-offs of these planktonic blooms led to the accumulation of protist shells on ocean floors forming extensive deposits in layers.

Limestone has been formed at the bottom of oceans by green and red algae and foraminiferans, while chalk consists of the coccoliths of golden algae. Reefs built by metazoans – mainly corals, but also sponges, mollusks and colonial Bryozoa have also contributed to the formation of limestone rocks.

Radiolarite chert is formed from the tests of radiolarian protozoans. I have proposed that bedded chert – which formerly may have been very extensive – was formed by diatoms which are motile golden-brown and yellow-green algae living as plankton in the oceans. This would mean that pure quartz (SiO_2) has been manufactured biologically. Being resistant to weathering, but prone to shattering, quartz mostly exists as sand incorporated into sandstones or as a component of other rocks.

Therefore, it appears that the sedimentary rock strata of the Earth were formed under aerobic conditions by aerobic forms of life. The first types of sedimentary rocks were formed as the seabed by the microbial transformation of basalt minerals, and the protistan deposition of other minerals. These processes continued on geological timescales.

The processes which raised these sedimentary strata above sea level to form land are explained in the Earth Shrinkage Hypothesis.

Granite

Granite is always classified as an igneous rock produced by magma – which it is, however, granite appears to contain a large contribution from sedimentary rocks. Molten magma rising up beneath sedimentary strata appears to have been transformed into granite by the melting and mixing of quartz contained within sedimentary strata such as bedded chert. The capping effect of the sedimentary strata has prevented the granite from reaching the surface, and so it has cooled as massive intrusions below surface.

Granite is a large component of landmasses – granite intrusions form the basement rock of many mountain ranges on land. Granite tends to only be exposed later on by weathering and erosion.

The Earth's crust became divided into landmasses composed mostly of sedimentary strata and granite, and ocean floors of basalt with a covering of marine sediments. According to the Earth Shrinkage Hypothesis landmasses may have changed in their extent, their height and their shape, but they have not migrated across the face of the Earth as claimed in the Theory of Continental Drift.

The **Earth Shrinkage Hypothesis** (Chapter 21) states that:

The Earth's core is cooling and shrinking in size, while the Earth's crust is essentially cold and resistant, and does not shrink. The Earth's hot mantle may adjust its size, while the crust is put under forces of compression. The crust adjusts to the smaller size of the interior regions of the Earth by the geological processes of folding, faulting and subduction.

According to the Earth Shrinkage Hypothesis the geological processes observed on the face of the Earth occur because of the Earth's crust *inability* to shrink. As the Earth's core has undergone progressive shrinkage, the surface area of Earth's crust has been reduced by it being folded up into mountain ranges; rent along fault lines with one slab of crust over-riding another slab of crust; and tucked in where crust disappears into the mantle at oceanic trenches.

The mantle has adjusted itself to the shrinking core by decompression causing it to melt in places and erupt upwards forming magma chambers which are emptied via volcanoes on the surface of the Earth. Volcanic eruptions move material from the mantle to the surface of the Earth providing an over-flow outlet.

It is the compression of the Earth's crust produced by shrinkage that has caused the continents to be pushed up above sea level, while reduction in surface area has caused ocean basins to be reduced in size – the extent of this is demonstrated by the actual disappearance of the Tethys Ocean in the location of the present Alpine-Himalayan mountain ranges. The once larger planet Earth with shallow seas now has deep oceans – due in part to the planet being smaller in size, and in part to an increasing amount of water on Earth produced by life.

Earth shrinkage also provides an explanation for the formation of mid-ocean ridges: Fault lines are found in the middle of oceans with transverse faulting dividing the oceanic crust into sections or slabs. Sections of crust can move relative to each other.

General Earth shrinkage causes the two parts either side of a fault line to be pushed together and rise up in the middle. The mantle underneath these fault lines decompresses and melts as the 'lid is lifted'. Then molten magma rises up along the fault line to form a basalt ridge.

Geodynamos and magnetic fields

The last chapter, Chapter 22, is about the magnetic fields of the Earth and the Sun. Earth's magnetic field emanating from the core is characterized by occasional polarity reversals over geological time; constant wavering in strength and an oscillation in direction; and is currently non-aligned with the axis of Earth's geographic poles by 11°.

The Sun's magnetic field emerges from sunspots; is strongest when solar flares occur; and it engulfs the solar system, flowing past the Earth as solar wind. The Earth is protected from the accelerated particles of the solar wind by its magnetosphere. The magnetosphere is formed where Earth's magnetic field traps particles from the solar wind forming a shield around the Earth.

I have proposed that the geodynamo of the Earth has two modes of operation: one mode can be likened to a direct-current generator with conducting fluid producing shifting electric currents and permanent magnets producing the field; the other mode of operation can be likened to that of a battery involving chemical energy and the transport of ions through a conducting fluid between electrodes. Reversals in the polarity of Earth's magnetic field would correspond to suddenly occurring changes in the mode of operation.

I concluded that the type of dynamo producing the magnetic field of stars such as the Sun, but also of brown dwarfs in which category I place the planets Jupiter, Saturn, Uranus and Neptune, would be different to Earth's dynamo mechanism. I have likened the dynamos producing magnetic fields around the Sun and gaseous planets to magnetohydrodynamic power generators (MHD) which function with plasma in the form of liquid hydrogen.

I have not presented these ideas concerning dynamos as another hypothesis, but as predictions of the Planet Capture Hypothesis and Earth Shrinkage Hypothesis.

The notion that the Earth's iron core was forged in a star allows the analogy of cast iron to be used in the search for understanding. The cooling of cast iron raises the possibility that a liquid iron core is surrounded by a thin skin of solid iron. (The current idea that the inner core of the Earth is solid is rejected). The notion of the shrinkage of Earth's core allows the idea that the core has shrunk away from the solid mantle leading to decompression in the lower layers of the mantle which has melted and become liquid magma.

Independence of core and mantle would explain the 11° non-alignment of Earth's magnetic field axis with the geographic axis. It could indicate that the core and mantle are rotating on different axes due to slippage. Rotation on different axes would produce friction between the core and mantle with the production of heat.

The toggling between two modes of operation in the generation of Earth's magnetic field could be explained by the forming of a solid skin around the molten core, followed by the increasing break-down of this skin in places allowing the influx of molten liquid mantle into the liquid core, and then the reforming of the solid iron skin. This would explain reversals in Earth's magnetic field.

The cast iron analogy and iron alloy composition of the core shows that the cooling iron core would form a solid skin around the outer edges with a 'mushy zone' below this. On the assumption that solid iron at high pressure in the interior of the Earth can be ferromagnetic, the permanent magnetism of the solid iron layer around the core and solid pancakes within a mushy zone would help to explain many features of Earth's magnetic field in terms of its wavering.

A further prediction of the Planet Capture Hypothesis involves the possibility that the Sun's dynamo mechanism is kept primed by rocky bodies captured from outside the solar system being engulfed by the Sun in the manner of 'seed' flung into an MHD generator to maintain its electrical conductivity.

Thus, I reiterate that four scientifically testable predictions can be drawn out from this new analysis of the problem of the origin of Earth's magnetic field and the Sun's magnetic field:

- Independence in the rotation of core and mantle. This is indicated in the 11° non-alignment of the Earth's magnetic field axis with the geographic axis.

- A solid outer layer of iron around the core which would play a pivotal role in the two dynamo modes of Earth's core. This would explain reversals in Earth's magnetic field.

- The existence of a 'mushy zone' of partial solidification causing wavering in Earth's magnetic field as permanent magnetism rises with solidification and falls with liquefaction.

- The engulfing of small bodies by the Sun as the cause of sunspots from which the Sun's magnetic field emerges.

These predictions are offered in support of the Planet Capture Hypothesis and Earth Shrinkage Hypothesis.

As it can be seen, the Planet Capture Hypothesis, Mass Extinctions on Earth Hypothesis and Earth Shrinkage Hypothesis support each other in providing an alternative view of the formation of the Earth and other planets. It has taken me many years to formulate these hypotheses and finally present them as a consistent picture.

Religious interpretations

Did God create the Earth as a planet with the right conditions for life to exist and create life on Earth?

Did God survey the heavens and choose Earth as a planet already possessing the right conditions for life as a good place to create life?

These are possibilities which I considered.

I reached the conclusion that life had created the conditions for life to exist on Earth. A succession of life forms – especially unicellular life forms – have transformed planet Earth from being a barren place into a biofriendly place with liquid water, an aerobic atmosphere and continents formed of sedimentary rocks and granite.

I concluded that the Earth was formed; it was not created.
Planets were formed by natural processes and became part of the solar system through the action of gravity exerted by the Sun. Likewise, other planetary systems have formed in a similar way around other stars.

Where did God create the first life?

I thought that life was probably not exclusively created on Earth. My first idea (on 15th December 1999) was that life may have been seeded everywhere in the universe where life is possible. If this were so, life was seeded on all the terrestrial planets, moons and even small bodies, but it only survives on a few of them.

Later I learned that water is abundant in space. If water is produced principally by microbial life, then there must be life in space. If water ice is associated with interstellar dust grains found in dark clouds, unicellular life possibly exists in these clouds.

I formed the idea that God created life throughout the universe in association with interstellar dust within the molecular clouds found within galaxies. If this were the case, the first microbial life on Earth found itself here by inoculation as the proto-planet passed through dark clouds of interstellar dust. Other planets and small bodies would also have picked up microbial life while travelling through interstellar clouds, and so microbial life or its former presence is likely to be found everywhere.

The Earth is always presented as an ideal place for life – life evolved here because it was a good place for evolution to take place. I present the view that it is, in fact, life itself that made the Earth a good place for life.

The Earth like other planets was formed by natural processes. I present the view that the Earth picked up microbial life and the ecological conditions produced by this life were allowed to develop naturally. Many planets underwent ecological disaster and lost the life they carried. The Earth's orbit in the inner solar system, the size of its molten core and its speed of rotation allowed life to survive on Earth.

As the Lord surveyed the heavens, there is a sense in which He could have chosen the Earth as a good place for the continuing creation of life since its ecology provided the conditions for subsequent acts of creation. There is another sense in which God created life in all possible places in the universe.

It may be that unicellular life continues to exist in interstellar dust clouds and on many planets, but others have lost the life they once carried. As far as higher forms of life are concerned, there is no information available on the possible development of multicellular life anywhere other than on Earth.

Concluding remarks

I have not included these Earth formation hypotheses in the Theory of Nanocreation and Entropic Evolution. Although I myself adhere to all the hypotheses presented in the volumes of this book, I am aware that my reader may choose to accept the Theory of Nanocreation and Entropic Evolution, while still believing in Plate Tectonics. That is a valid position. There are other readers who will take up some of the hypotheses of the Theory of Nanocreation and Entropic Evolution, while rejecting others. Still others may be interested in these hypotheses involving natural processes while rejecting notions of creation.

My hope is that dialogue and freedom of thought will move things forward, after all, what we all care about is science. For my part I seek more than this – in the presentation of these hypotheses I seek the glory of God, and that He may be worshipped as the Creator of the universe, life and ourselves.

APPENDICES

Shrinkage in iron

Expansion and contraction with increasing or decreasing temperature in iron is cubic – this means that it is equal in all directions. However, the expansion of iron is usually given as a coefficient of linear expansion. This is because it is arrived at by measuring metal wire and most engineers are interested in the expansion of iron in the solid state.

Liquid molten iron
Nearly all substances have greater volume in liquid state than in solid state. This is true of molten rocks and metals including iron. (It is not true of water: solid ice takes up a greater volume than liquid water). Liquids increase in volume as the temperature goes up – an example of this is the thermometer filled with liquid mercury. There are also thermometers filled with gas which are used to measure very low temperatures.

Iron alloys undergo a phase change between the liquidus temperature and the solidus temperature. In molten liquid state the atoms are unbound, whereas in the solid state the metal atoms are bound into the lattices of crystals. The liquidus and solidus temperatures are different for each iron alloy depending upon its composition.

The liquid contraction of iron from the superheat temperature to the liquidus temperature is very predictable – it is dependent upon the coefficient of expansion of the alloy. This is generally around 1.5 % by volume per 100 $^{\circ}$C.

Solid iron or steel

Liquid shrinkage through the liquidus temperature involving a phase change from liquid to solid is uneven and depends on temperature.

When iron becomes a solid, it forms cubic crystal structures which may have a body-centred cubic structure with an iron atom at each corner of the cube and an extra atom in the middle or a face-centred cubic structure with an iron atom at each corner and an extra atom in the middle of each face of the cube. Body-centred crystals have a higher specific volume than face-centred crystals.

When hot liquid iron cools down to 1538 °C, it contracts as it forms delta-ferrite with a body-centred cubic lattice structure. When it cools down further to between 1394 °C and 912 °C, it forms austenite with a face-centred cubic lattice structure. The face-centred crystal structure occupies a smaller volume than the body-centred structure, and so the iron undergoes greater shrinkage at this stage. Below 912 °C alpha-ferrite is formed which reverts to a body-centred lattice structure with higher volume, thus a slight expansion occurs.

When steel tools are made they are hardened by the process of quenching. This involves the rapid cooling of austenitic iron alloys faster than the critical cooling rate. Quenching prevents diffusion from occurring such that a body-centred tetragonal lattice of iron atoms forms with a carbon atom wedged inside it – the carbon remains in solid solution. This strongly distorted crystal is called martensite and it starts to form below 240 °C. Martensite forms in the steel tool which has been sent for hardening. Martensite has a larger volume than austenite and alpha-ferrite and so produces a relative expansion in this part of the steel. Depending on the content of the steel alloy the increase in volume with martensite formation may be between 0.1 and 1 %.

The overall solid contraction of iron alloys is dependent on the expansion coefficient which is measured linearly on metal wire at different temperatures.

The makers of cast iron or steel tools and components are concerned with the contraction of iron upon cooling. Uneven contraction can cause cracking in the finished component. Micro-shrinkage leads to the problem of porosity in the metal product. Iron casting processes use moulds designed for the continuous feeding of molten metal into the casting to keep it full as cooling and solidification proceeds with the concomitant shrinkage of iron.

In the making of steel tools, macro-shrinkage is always allowed for by making the mould larger than the desired dimensions of the component or machine tool.

NB The temperatures given apply to the Earth's surface with one atmosphere. The atmospheric pressure of the Earth's inner core is 3.3-3.6 million atmospheres.

Earth's core

As far as the Earth's core is concerned, it has the composition of an iron alloy possibly containing 4-5 % nickel, 1.5 % chromium, 2-4 % carbon, 2-4 % iron carbide (Fe_3C) and >2 % silicon.

I am assuming that the Earth's core is entirely molten liquid except for an outer solid skin as explained in Chapter 18, section 1 and Chapter 22, section 5. The Earth's core is losing heat at a known rate as infrared radiation from the surface of the planet into space (see Chapter 18, section 3). The liquid contraction of iron alloys is around 1.5 % by volume per 100 °C. The volume of the liquid core and the rate of heat loss would allow one to calculate the rate of shrinkage occurring in the Earth's core. A solid peripheral part of the core would have undergone greater shrinkage as it was transformed from a liquid to solid state. Thus, the existence of a solid part of the core would accentuate shrinkage of the interior of the Earth.

Bibliography

Bohler Bros. & Co. Ltd. Manufacturers of High Grade Steel *Bohler Hints for the Hardening Shop*

www.ductile.org
Ductile Iron News Issue no.3 2001 Article: Shrinkage in Nodular Iron

Other iron casting websites include:
www.itm.ac.com
www.locknstitch.com

Encyclopedia Britannica 2011 Standard Edition: Iron processing

ABOUT THE AUTHOR

The author studied anthropology and sociology at Oxford Brookes University in England and Aix-en-Provence University in France obtaining a BSc Honours degree in 1984, a 'Licence' in 1988, and a 'Maîtrise' (masters degree) in 1990.

She is a practicing Christian attending the local Roman Catholic Church.

Lightning Source UK Ltd.
Milton Keynes UK
UKOW04f1004260816

281560UK00005B/16/P